T0340204

A Primer in Financial
Data Management

A Primer in Financial Data Management

Martijn Groot

ACADEMIC PRESS

An imprint of Elsevier

Academic Press is an imprint of Elsevier
125 London Wall, London EC2Y 5AS, United Kingdom
525 B Street, Suite 1800, San Diego, CA 92101-4495, United States
50 Hampshire Street, 5th Floor, Cambridge, MA 02139, United States
The Boulevard, Langford Lane, Kidlington, Oxford OX5 1GB, United Kingdom

Copyright © 2017 Elsevier Ltd. All rights reserved.

No part of this publication may be reproduced or transmitted in any form or by any means, electronic or mechanical, including photocopying, recording, or any information storage and retrieval
system, without permission in writing from the publisher. Details on how to seek permission, further
information about the Publisher's permissions policies and our arrangements with organizations
such as the Copyright Clearance Center and the Copyright Licensing Agency, can be found at our
website: www.elsevier.com/permissions.

This book and the individual contributions contained in it are protected under copyright by the
Publisher (other than as may be noted herein).

Notices
Knowledge and best practice in this field are constantly changing. As new research and experience
broaden our understanding, changes in research methods, professional practices, or medical treatment may become necessary.

Practitioners and researchers must always rely on their own experience and knowledge in evaluat-
ing and using any information, methods, compounds, or experiments described herein. In using such
information or methods they should be mindful of their own safety and the safety of others, including
parties for whom they have a professional responsibility.

To the fullest extent of the law, neither the Publisher nor the authors, contributors, or editors, assume
any liability for any injury and/or damage to persons or property as a matter of products liability,
negligence or otherwise, or from any use or operation of any methods, products, instructions, or
ideas contained in the material herein.

Library of Congress Cataloging-in-Publication Data
A catalog record for this book is available from the Library of Congress

British Library Cataloguing-in-Publication Data
A catalogue record for this book is available from the British Library

ISBN: 978-0-12-809776-2

For information on all Academic Press publications visit our website at
https://www.elsevier.com/books-and-journals

Working together
to grow libraries in
developing countries

www.elsevier.com • www.bookaid.org

Publisher: Candice Janco
Acquisition Editor: J. Scott Bentley
Editorial Project Manager: Susan Ikeda
Production Project Manager: Nicky Carter
Designer: Mark Rogers

Typeset by Thomson Digital

Contents

3.   Information as the Fuel for Financial Services' Business Processes

6.   Data Management Processes and Quality Management

Foreword

The Financial Services industry has been characterized by a high relative spending on systems and software technologies since these became available. According to most industry surveys, per capita spending on technology is considerably higher than in other industry segments, in some cases double or more.

Given that financial products lack "manufacturing friction"—there is no assembly line waiting for pre-manufactured components to be bolted together—the "arms race" for competitive advantage focuses on information advantages. But the dirty little secret of the industry is also tied to this lack of friction. Because markets absorb information quickly, any product, investment strategy, or service which has a true competitive edge derived from information or execution characteristics, or a combination of both, will lose that edge in a short timeframe, as other market participants react to the "new" competitive requirements.

So we have an industry landscape characterized by accelerating rates of change and adaption in terms of product offerings coupled with the same type of change profile in the mechanics by which these offerings are implemented—systems technology. These two factors are synergistic—ever increasing complexity in instrument design and risk analytics are fed by increased availability of large data sets tied together in previously unworkable ways.

For practitioners in the Market Data world, whose mission in life is to make sure that the data is clean, reliable, accurate, and accessible from almost any conceivable perspective, there are other relevant factors as well. First, regulators are scrambling to keep up with and anticipate market developments. The complexity of regulatory requirements, especially across multiple jurisdictional boundaries, is a significant business issue for all involved. Next, the nature of the business requires multiple information sources and methods for combining them into a coherent view depending on internal or external customers' requirements. Here we run into the industry's dirty little secret, one that you won't find discussed much in the vast majority of books or articles about or for Financial Services practitioners. The secret is relatively simple:

As a general rule, every system that was ever used to address a business process in a bank, brokerage firm, investment firm, or any of the related constellations of businesses that make up what we call the Financial Services Industry, is still out there. Either that piece of code or that piece of hardware (or both!) are still being used in operations, or the "new" system designed to replace it is shaped by how the first one was built.

This complexity is a serious issue, especially for those tasked with maintaining, or migrating and moving forward these systems, and basically unaddressed in the existing literature until now; and are spotty at best. The publication in 1986 of the first version of David Weiss' book "After the Trade Is Made: Processing Securities Transactions" has been followed after a long silence by a plethora of *new books touching on and around securities processing mechanics, but none suf-ficient as a guidance framework for those who do it for a living. Realistically, the field is changing too fast and both the data and the methods for dealing with the data ensure that most of the really good stuff stays locked up in training materials within firms. For the most part, by the time the material is no longer considered proprietary, it's out of date.*

A corollary of the rapid rate of change in practice is the lack of a generally accepted framework which can serve as a "coordinate system" for industry dis-cussions. Everyone talks about their piece of the puzzle, but putting together a consistent prototype of how different actors, their actions, competing goals, and interdependencies interrelate is something the industry has not yet sorted out.

Which brings us Martijn Groot's "Managing Financial Information in the Trade Lifecycle: A Concise Atlas of Financial Instruments and Processes".

The title itself is pretty ambitious. Fortunately, Martijn pulls it off. He also goes a long way towards solving the framework issue. His approach is based on a combination of viewpoints, intersecting instrument and transaction lifecycles, and tying them together within a Supply Chain model borrowed (at least origi-nally) from advances in Manufacturing and Operations Management. While not the only industry expert attracted to the Supply Chain approach, Martijn has done a masterful job of adapting it to the realities of Market Data practices. Here, for the first time, is a coherent descriptive framework that describes how the pieces fit together, why they need to be handled in the way they are and what metrics and aspects of information quality can be used. It doesn't go into the mechanics of pricing and valuation for various instruments—that's not the point here, and there are plenty of sources for information on that. There are other places where Martijn neatly draws the line, and stays focused. If you need a concise overview of information management and the products and processes in the operations of the securities industry, including practical discussions related to trade-offs and legacy overhang, you now have it.

Bill Nichols
Practice Manager
Technology and Standards
FISD 2007

The foreword cited previously was written for this book's predecessor almost a decade ago. The financial services industry has since experienced significant shocks, from the global Credit Crisis to the G20 calls for transparency in OTC markets and the accompanying regulatory expansion within and across borders. Unprecedented bailouts of large banks have led to a smaller number of larger firms, and the resulting "barbell" concentration of firms and flows is still evolv-ing. The initial spate of analysis and regulation that seemingly inevitably fol-lows market breakdowns is gradually evolving into a more thorough synthesis. There is a growing awareness of the extent to which exponential growth in the capabilities of underlying horizontal technologies is creating new possibili-ties—in product and services, markets, firms, and organizational and operating approaches within firms.

One of the "long tails" of the financial crisis was a heightened awareness of the importance of well-managed data in general and especially in times of stress. Some very basic questions—"Can you get the right data when you need it? At the right level of granularity, and with measurable precision?"—went unanswered at the worst moments. Firms didn't understand their own risks or obligations well enough, and regulators had little if any information about market structures or products that were radically mispriced or traded in opaque and brittle markets where apparently deep liquidity vanished in microseconds.

Data is the name of the game today. How well we understand the nature of the game is still open to debate. At a minimum, we know that market reforms and new or expanded regulatory requirements have demonstrated how deeply interconnected and complex markets have become—and are increasingly becoming. The effort to create globally harmonized Trade Repositories for OTC derivatives, for instance, has brought into sharp relief the complexities involved—markets are global, products are complex, and the window for decision making is ever smaller. The underlying horizontal technologies are evolving at breakneck speeds, while vertical, financial services–focused technologies and tools are becoming increasingly specialized. The disparity between investment in the front office versus that in the middle or back office and the accompanying multiplier effect on "data density" for each incremental level of specialization has put further pressure on information management.

According to a marketing paper from IBM released in December 2016, "90% of the data in the world today has been created in the last two years alone, at 2.5 quintillion bytes of data a day ...." Needless to say, the technological capability to create and manage data at this rate of growth didn't exist 10 years ago, or realistically even 5 years ago. This highlights another critical data issue facing both industry and regulators: Simple throughput—where do you put the data and how fast can you crunch it to get the needed results?

Most large firms have multiple legacy systems with roots going back decades, with estimates of 75% or more of systems spend dedicated to maintenance of existing systems. This cannot continue—the exponential expansion in data creation is starting to yield volumes of data that simply can't be managed by systems operating under legacy constraints. We are now seeing different types of emergent operational risk that require a more disciplined understanding of and approach to operational management in financial firms. Among other things, there is a small but growing body of research on complex systems that includes studying some of the largest and most complex engineering projects ever created—financial markets.

More than large logistics networks such as those in retail or manufacturing, the composition of each part of the financial services system—Buy- and Sell-Side market participants, exchanges and market networks, payment systems, and clearing—has significant variability across regulatory reporting requirements, standardization, and technology and data management practices, to name just a few dimensions. There is some evidence that connecting and integrating the disparate parts of current market practices and their supporting infrastructures leads to unpredictable responses—but at this point it's not clear how we would detect, let alone quantify, aberrant information resulting from the interactions of many complex systems connected to each other. And to be clear, we are not yet talking about the semantics of the data in these systems: what it means, how it got there, why it got there, and what it is for. The "unpredictable responses" cited previously come from automation attempts across different firms' infrastructures. The idea that simply connecting applications across market participants, plugging them in, and turning them on creates the potential for intermittent and inconsistent "noise" is not a comforting one, but at least for the moment we can leave this one for researchers in the computer sciences. It is important, however, to remember the operational risk in building our data systems on increasingly complex and in some sense "unknowable" underlying landscapes.

Across multiple industries, practices are emerging that address these and other issues related to expanding information capabilities and requirements— better ways to "change the tires while driving the car." Increasingly, the operational requirements for managing data are going to require firms to make new choices in the way they manage systems resources—choices arising from how you manage to kill off legacy systems smoothly while increasing capacity. Those that are best at this are likely to enjoy temporary but significant competitive advantages. This will come only through a rethinking of the value chain(s) within which financial firms compete.

Given the related investments from the public and private sectors, one would hope that the quality of information in the industry wouldn't be an issue. The melding of technology and financial markets is irreversible and accelerating, and management thinking and operational practices need to keep up. This can be done well only if informed by practical expertise based on experience.

Which brings us to this book. As noted earlier, the title of Martijn's previous work is 15 words long, and the text drills deeply into some of the related operational details. The focus is on practitioners and is generally oriented to a department-level view. The title of this book contains six words. It covers a much broader range of issues and addresses two of the most important concepts that have moved to the fore of late—information governance and new operating and organizational possibilities resulting from systems evolution. There is a much wider perspective examining some of the interplay of market trends, regulatory responses, political realities, and related technology developments.

Martijn is ever a practitioner's practitioner, with the accompanying grounding in pragmatic organization of thought. The book's breadth and depth of content about what it means to operate, manage, and evolve financial systems, the discussion of industry and regulatory approaches with regard to shared utilities and distributed transaction registers, and the recognition of how consumer technologies are affecting expectations and capabilities across every industry, but especially finance, all are part of a pragmatist's responses to the expansion of the domain and range of requirements with which those "on the ground" today in the financial industry must deal with.

Compared to 10 years ago, there are many more books in this space to choose from. For the most part, these are dedicated to either detailed operational or financial operations oriented around asset-class specifics. Martijn has taken a very different approach here. This is both a "big picture" and a "be organized" perspective. If you work in the industry, or want to understand some of the key issues faced by those who do, you will recognize a few or even many of the topics presented here. But the integration and formalization of the framework that is developed throughout the book is profoundly, practically, useful to anyone in the industry faced with the task of actually doing (and staying on top of!) the work required to manage **all** of the information for a financial firm.

In short, this a working example of how to think about the ever-expanding number of factors, developments, practices, and changes—what they mean, and how they relate to each other—necessary to swap out your car's tires while driving 120 km/h at night on a wet winding road with one headlight out. If you're skilled and experienced, pragmatic and open to new ideas and approaches, you can probably figure out how to put a new bulb in the headlight at the same time. Good luck!

Bill Nichols
Senior Advisor
Information Architecture and Innovation
OFR 2017

Preface

In 2007/08 I wrote *Managing Financial Information in the Trade Lifecycle* that looked at financial information management from a supply chain perspective. Given the rapid changes since then in business, customer, and regulatory demands as well as the developments in information management and enabling technology, I felt it was time for an update. Because I also wanted to put these new developments and requirements in a broader context of financial services' business processes beyond trade lifecycle management, this turned out to become a rewrite rather than an update. This book is the result.

Martijn Groot

Chapter 1

The Changing Financial Services Landscape

Chapter Outline

1.1 DATA AS THE LIFEBLOOD OF THE INDUSTRY

The book gives an overview of the challenges in content management in the financial services industry. It is both an update and an extended version of a book I wrote back in 2007 just before the onset of the financial crisis: *Managing Financial Information in the Trade Lifecycle: A Concise Atlas of Financial Instruments and Processes*. The current book differs in two important ways:

- Since the 2007–09 global financial crisis, business models of financial services firms have undergone enormous change and regulatory intervention and regulatory information requirements have significantly increased.
- The technological drivers for change have accelerated and—if a crisis and regulatory scrutiny were not enough—the financial services industry is also challenged by disruptive new entrants. Customer expectations on interaction with their financial services suppliers pushes firms to change.

In other words, an updated version is in order, a version that takes the notion of a *Primer* as a starting point: back to first principles when it comes to information management in financial services. What do regulatory intervention and

A Primer in Financial Data Management. http://dx.doi.org/10.1016/B978-0-12-809776-2.00001-6
Copyright © 2017 Elsevier Ltd. All rights reserved.
1

common regulatory themes, such as solvency, liquidity, investor protection, and pre- and posttrade transparency in OTC markets mean from a financial services information perspective? What do customer interaction expectations mean for the back-end infrastructure? What does the move to the cloud and mobile interaction mean for security and for the information supply chain? How can financial services firms innovate and capitalize on new technology?

These are some of the questions we will be exploring in this book. We will discuss best practices and recommendations on information management **seen from the data perspective**. A financial institution and increasingly any kind of business can be seen as a collection of data stores and processes to manipulate that data and to bring new data in as well as to push data out—to regulators, investors, business counterparties, and customers. If we see the financial services industry as a network consisting of **actors** (clients, banks, investment management firms) and **transactions** (account opening, money transfers, securities transactions) between these actors. We can see business processes from the perspective of transaction life cycles—research, trades, and posttrade activities—as well as master data, changes, such as product and customer lifecycle management.

No other industry is as information hungry as financial services—all the raw material is information itself. More than in other industries, capabilities in information management are more important. The potential impact of the financial services industry (especially the adverse impact) on the real economy has been well documented (see, e.g., United Nations Environment Programme, 2015). The irony in financial services is that this is an industry where the need for information at the point of buying is largest—given the length of some of these products (life insurance, mortgages) and the far-reaching impact they can have. The far-reaching impact of financial products buying decisions for consumers (insurance, investment/retirement plans, and mortgages) contrasts with the relative ease by which these products are marketed and bought.

Information and timing is critical both in wholesale banking and in retail banking due to the speed of technological innovation. The large amounts of additional data generated and the different ways in which customers transact with their financial service provider have led to new demands on information technology, information availability, and security.

In this introductory chapter we will discuss some of the recent developments in data management. This will be followed by an overview of the supply chain perspective in information management—seeing it as a *logistics* problem. We will end this introduction by stating the various aspects of the data management problem to set the stage for the next chapters.

The reach of the book is broad so necessarily some topics will be discussed at an introductory level and some areas will be explored more in depth. Focal areas are information management from a process perspective and how data management considerations differ by the type of information and its use cases.

1.2   DEVELOPMENTS IN INFORMATION MANAGEMENT

Data management has come on the radar in recent years since its successful rebranding into "big data." Big data is nothing more than the application of today's information aggregation tooling and hardware processing capacities to business processes—ranging from upsell suggestions to call center staff to credit scoring to uncovering investment strategies. The main developments that have made data management more critical than ever in financial services are as follows:

- Growth in the volumes of information. Customers interact using mobile devices and leave an extensive digital trail.
- Faster transaction and settlement cycles shown by the advent of high-frequency trading and shrinking settlement windows.
- Speed of technological innovation and the competitive changes introduced by those. Computing power has increased and technologies created and brought to fruition by internet retail companies and social media start to become applied in financial services.
- Regulatory information and process demands. Regulators ask a lot more detail and since regulatory reporting is a central function, this is where the onus is on connecting different internal information sets that are typically scattered by customer segment or product verticals. Simultaneously, regulators scrutiny the quality of internal processes and quality metrics.
- Less tolerance and more demands on interaction from customers. Financial services are no longer a "special" service. Used to other retail services provided over the internet, clients expect high standards when it comes to their account overview, order status, and response times. This puts pressure on the back-end infrastructure and information aggregation capacities of banks.

To start, let's look at the growing volumes of information. Traditionally, in data management the focus of volume growth had been in the wholesale markets. Rapid economic developments in certain areas of the world, a move to on-exchange trading and more trading venues—as well as growth in the number of hedge funds and the rise of high-frequency trading—**all led to more transactions**. To give some idea, large exchanges have a daily volume in the millions of trades (see https://www.nyse.com/data/transactions-statistics-data-library), central securities depositories clear in the hundreds of millions, and swap trades may be in the single millions (see https://www.euroclear.com/dam/PDFs/Corporate/Euroclear-Credentials.pdf for statistics; see http://www.swapclear.com/what/clearing-volumes.html).

Postfinancial crisis, the growth in available information on retail and SMEs is perhaps more important. Due to mobile interaction and the online presence of consumers and companies, the amount of available information to be used in credit scoring, prospecting, and upsell decisions has exploded. Customers, often inadvertently, leave a lot of information.

The *lag* between the moment of the transaction and the moment of settlement is shrinking. A lengthy settlement time brings operational risk into the process. The longer this lag, the larger is the potential outstanding balance between counterparties and the higher the settlement risk. At the same time, regulations, such as Dodd–Frank in the United States and EMIR in the European Union have pushed product types, such as interest rate swaps that were cleared bilaterally to central clearing. This means information needs to be available faster and the time available for error correction is lower.

Hand in hand with the volume developments are the available technologies to act on these new information sets. Recent developments in hardware have lowered the cost of storage and of computing power. On the software side there are many more tools that access data—so the cost of manipulating data has become lower.

The advent of Web 2.0 and social media have pushed a revolution in data storage and access technologies. The introduction of NoSQL and other non-traditional database technologies made for cheap ways to achieve horizontal scaling—which offers ways of handling and processing much larger sets of information. Historically, data needed to undergo an elaborate curation process before it could be used to feed analytics. New ETL (ETL stands for Extract, Transform, and Load) and analysis tools will absorb whatever data they can and get cracking. This is potentially dangerous as data may be misinterpreted or ignored without the user drawing on the resulting statistics being aware of this.

Traditionally banks had lengthy information curation processes—by which data was sourced from branch or call center interaction and manually entered into predesigned data templates. This was the era of structured data. Now clients create large amounts of **unstructured information,** such as social media posts, emails, and news as opposed to **structured information** that lives in fixed data models. The ability to process this unstructured information leads to new possibilities. More data is available and more data can feed into bank's processes. An important result of technological innovation is that the point at which data can start to feed into analytical, decision-making processes has moved upstream from curated content to directly work on the raw materials to draw inferences on customer buying preferences (Fig. 1.1).

At the same time, the fact that internet companies, such as Facebook, Google, Apple, and LinkedIn occupy a prominent place on customers' smartphones meant they could become the front door of a larger consumer mall, the main access point to financial services, and insert themselves between customers and the traditional banks. The battle for the desktop or phone screen is ongoing and a banking license is easily acquired. An additional challenge is the public image of banks versus the public image of internet companies. Following the crisis it has become harder for banks to position themselves as "trusted third parties" that should necessarily be a part of transaction chain. Especially on the payments side the rapid evolution in ecommerce has introduced many new fintech companies that provide services that banks could have provided.

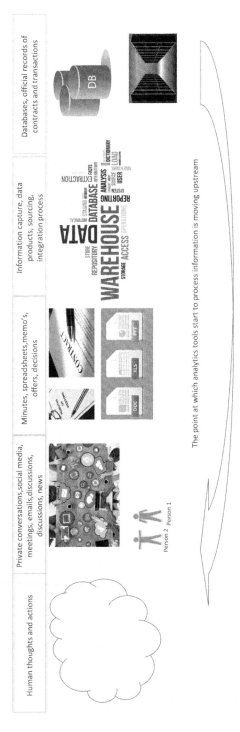

FIGURE 1.1 Data curation process changes.

1.2.1 Regulatory Demands

The regulatory response to the crisis has been far-reaching and has accelerated existing trends in data aggregation capabilities. Regulatory policy objectives include:

1. strengthening banks' balance sheets by increasing the capital buffer to make them more resilient;
2. making markets more transparent by promoting pre- and posttrade transparency;
3. protecting the general public from bad financial products;
4. instilling a cultural change at banks to make them more risk aware and to focus not just on risk models but also on the entire process.

By definition, regulatory reporting is a central function. Central banks or securities markets regulators need the global picture and the risk and finance functions are typically the places where all the different threads come together and where the data integration problem is most acute. New regulation on solvency comes down to providing more detail on the exposures, additional risk categories that were not previously separately reported on, new metrics to capture tail risk, and "what if" scenario analysis in the form of stress-testing exercises. From a transparency perspective, new regulation introduced obligations to report transactions and to standardize the information provision of financial products. Interestingly, this has led to a lot of raw material in the form of published transactions available that can be used in market risk and valuation processes.

A focus on investor protection has caused banks to collect additional information on all their clients to screen their investment practices and to determine their eligibility to trade in different asset classes. Before the crisis, risk seemed to have become a continuum that could be sliced, diced, blended, and bundled. Financial products could be "built to order" for counterparties. This also meant that risk ended up where it was least understood. Retail, SME and local government misselling scandals testify to that. Consequently, he information reporting requirements on financial products and target clients alike have gone up.

Regulators have increasingly occupied themselves at microlevel in the identification and classification of financial products and legal entities that interact with them. One of the most successful examples has been the introduction of a global standard for the identification of legal entities, the legal entity identifier (LEI). Separately, jurisdictions have become more aggressive in chasing missed tax revenue—this has led to additional information that needs to be more accurately measured.

Finally, regulators have broadened their focus to look not just at risk models but also at the quality of the data and the quality of the processes feeding their models. The Basel Committee document BCBS239 lays down 14 principles for risk data aggregation that impact data management, which we will cover in Chapter 3.

The volume of regulatory documentation and corresponding IT costs have risen sharply in the past decade. (See http://www.chartis-research.com/research/reports/global-risk-it-expenditure-in-financial-services-2016 who estimate $70 billion global spend on risk IT systems and services. IDC has a similar estimate and expects spend on risk information technology and services to grow to $96.3 billion by 2018. Thomson Reuters estimates 8704 regulatory alerts in 2008 that grew to over 43,000 regulatory alerts in 2015. Source: www.reuters.com; a regulatory alert is a new document, analysis, or policy news item.)

1.2.2   Customer Preferences

The ways customers interact with their financial services firms is a very different driver—branch visits have become rare. Banks have started this by pushing clients to call centers—the unintended side effect was that financial services providers became increasingly interchangeable. The internet shopping retail experience makes for high expectations on service, returns, flexibility, availability, feedback mechanisms and a clear account overview.This has put pressure on the back-end infrastructure and information aggregation capacities of banks.

Most back-ends live still in the EOD, batch-oriented world. They are not equipped for 24/7/365 processing but instead reflect a single-site office working day rather than a globally connected infrastructure that is continuously up consumers ready to transact business via mobile apps). Because often technology from the 1970s, 1980s, 1990s, and 2000s coexists, there is poor integration between systems with a lot of potential for confusion at the handover points.

Especially for diversified banks that offer current account, loans, mortgages, brokerage, cards, and insurance the need to integrate all products sold to a single account is a major challenge as the back-end is often very fragmented. Customers, however, expect instant access to all products sourced from a certain firm. The challenges for financial services—apart from the integration of fragmented heaps of information—are regulatory and information security and privacy questions, which are much more pronounced than in any other retail shopping service.

Taken together, these developments in data management put pressure on information management capabilities. The eroding margins in banking have caused firms to be conservative in their IT spend when there is a hurdle to overcome. The good news is that data standards, integration tooling, and storage and access technologies have made huge strides in the past decade. The ability of the IT department to respond to these challenges is going to be one of the major competitive differentiators in the new era of financial services.

1.3 THE SUPPLY CHAIN VIEW OF DATA MANAGEMENT

In this section we discuss the supply chain view of information management that we'll use in the rest of the book. We start with an ultrashort history of automation in the financial services industry.

1.3.1 Ultrashort History of Automation in Financial Services

The first wave of automation in the 1960s was aimed at cheaply processing standard products. These products were mostly aimed at solving investment problems: that is, companies that needed money through equity or debt funding. At that time, the market for derivatives was tiny or nonexistent. With greater volatility and instability of the financial system came the need to *price* and subsequently *trade and transfer* various types of financial risk. This caused a flurry of financial product innovation.

When new product lines, such as interest rate swaps began to appear in the 1980s, this led to architectures of local (also called *decentral*) systems around the prevailing mainframes. Different applications with separate business logic were needed to accommodate the proliferation of products. However, it became apparent that the content needed to operate these applications suffered from this very decentralization. All too often information that enters a processing application arrives late or is faulty or incomplete.

Discrepancies in input cause havoc downstream when it comes to reconciling transactions and when aggregate financial reports need to be made. We will discuss the issue in more detail later, but part of the problem is the blending of information and business logic. Each application has its own storage that often independently sources information. Added to this is that all too often these stores have been tweaked to allow for variations in the products they are supposed to handle.

On top of that, many applications called end user–developed applications (EUDA) sit at the desk level. They were designed as *end points* of data streams to function as a local analysis tool or as a way to create and format reports. However, often these EUDA have taken on production system positions—leading to numerous issues in permissioning, security, audit robustness, and version control. To counter this, tools in middleware, Excel version control, and verification have been created. Excel should be a real EUDA; it should be the end point of a data stream, a presentation medium for an end user to communicate. When it functions as an intermediate part of a data flow that doubles as a staging tool, or a publishing tool into the next information processing application it can have disastrous consequences.

The consequences of a half century of automation are that organizational processes are locked in and hardwired into IT systems. IT speeds up existing processes but can also cement them and freeze them in time. The typical information architecture of a bank or investment managerlooks like something I call **the data**

swamp. Many streams of information come in at various locations in a process, sometimes multiple times. Multiple conversions take place as content needs to be moved from one processing place to another. Reconciliation applications have been put in place with no other function than to control the damage. Excel more often than not plays the role of superglue. Occasionally, attempts have been made to bring order to this jungle by putting in middleware or a data backbone. Typically political or technological problems have prevented not only the completion of these initiatives but even their fruition to a stage where they actually reduce some of the chaos (Fig. 1.2).

A large financial services institution may have thousands of applications. This means that the great bulk of the IT budget goes toward maintaining applications that often use obsolete technologies and that no longer fulfill the user's requirements. Worse, most of the applications integrate on a peer-to-peer basis leading to a vast amount of connections and potential failure points.

In a typical manufacturing business, the amount of waste found in financial services IT would be unacceptable. In many cases, the fact that things have not broken down completely is testimony only to the power of Excel and the

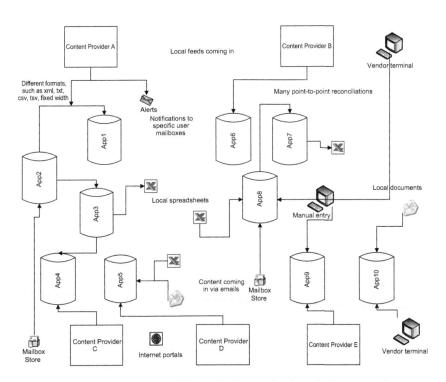

Local stores, embedded data models, local feeds, manual entry, duplication, and need to cross-reference and reconcile

FIGURE 1.2   Typical spaghetti architecture.

	1960s	1970s	1980s	1990s	2000s	2010s
Vendor products	IBM mainframes , vertical integration	Arrival of alternative platforms	Growth in capital markets fintech	Capital markets product silos	Outsourcing/ offshoring IT and ops	Shared services, cut in app numbers
Business Drivers		Floating currencies	Deregulation of capital markets	OTC derivatives boom	Basel II, hedge fund explosion	Cost reduction, widespread reform, regulatory change management
Financial Products	Stocks and bonds	MBS, currency options, equity options, financial futures	Interest rate swaps, CMOs	More exotics , credit derivatives	Emerging markets , exotic products	Back to basics, user experience, fintech competitors
Technology	Start of FSI automation		"Decentral systems "	Internet, middleware	Mobile breakthrough, banking apps , low latency race	Big data technologies , retail consumer fintech

FIGURE 1.3 Financial services information management history overview.

ingenuity of countless ad hoc reconciliation actions by desktop users (Fig. 1.3). Frequently, content which—when integrated—would be the basis of informed trading and investment decisions lies scattered across the firm. Why does this situation of inaccessible and fragmented data persist? Why is data often buried and inaccessible? How can its value be unleashed?

Practitioners in financial services know that the data issues all too well. Some of the top complaints include the lack of flexibility in searching information or introducing new data elements, different standards, a long lag in requesting and receiving report, difficulties in accessing data, arcane formats, lack of information on audit or lineage, difficulties in seeing historical information, and—above all—**constant** discrepancies.

The costs of this situation are threefold:

- *Direct recurring costs.* These costs help in maintaining and reconciling separate information flows, employing staff to keep track of them and to operate redundant products and services.
- *Costs through higher operational risks.* These costs vary from failed transactions and subsequent claims to higher capital charges for increased operational risk.
- *Opportunity costs.* Paradoxically, the spaghetti situation that arose partly because of time-to-market pressures means that time to market for new products is very long. Especially now with the availability of disruptive technology to new entrants, the opportunity cost is enormous.
- *Regulatory risk.* Regulatory scrutiny on processes means the number of process points that can lead to qualitative failures is very large.

Cloud has already changed the way data is archived and kept for consumers and businesses alike. New technologies in storage, integration, and analysis, such as NoSQL, business intelligence, and visualization tooling provide powerful options to change both infrastructure and end-user productivity. Technologies, such as distributed ledger will potentially disrupt processes that are deeply anchored in financial market infrastructure: in exchanges, clearing houses, and central securities depositories.

It is helpful to view data management as a logistical challenge and to view the industry as a set of actors and links between them. The interactions between actors demand or create information. The network nodes itself are instruments, counterparties who interact via trades and other contracts (Fig. 1.4).

Every action initiated, from the opening of an account to the onboarding of a new client, the creation and offering of a new product, or the trading of stock, is a decision. Every decision requires information that is delivered through an information supply chain that transforms content as it flows through various stages where it is sourced, verified, enriched, molded, analyzed, and finally acted upon.

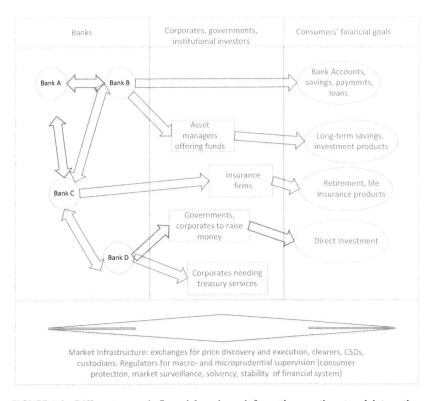

FIGURE 1.4 Different actors in financial services—information creation at each interaction.

1.3.2   The Information Supply Chain

The network of financial services is characterized by continuous change and the actors (clients, counterparties) come and go. Before a client is onboarded, compliance and credit checks are required. After that, regular checks and monitoring are required. New products are created: companies list, issue bonds, and investment structures are created and tuned to fiscal regulatory boundary conditions and to investor risk/return and horizon expectations.

The lifecycle management issues in instrument and clients may intertwine: pretrade research often leads to the creation of custom instruments for specific investors or investor groups. In the case of OTC derivatives, the start of the transaction and instrument life cycle coincide since every trade is a new, unique bilateral contract. The information industry services the needs of decision making at every stage. Behind the provision of information (research, streaming data, and corporate actions) to people and applications that need to act on it, there is often a lengthy supply chain with content being enriched, derived, bundled, and validated at different stages.

Traditionally, clients came to the infrastructure, first to physical branches, and then this moved to call centers and home computers; now the IT of banks stretches into every room, and every device; the bank comes to you rather than the other way round. The expectation of uptime anytime and instant access to aggregated views puts pressure on information that needs to be managed against a large set of boundary conditions on security, privacy, and regulation.

The supply chain of information (Fig. 1.5) has tended to increasing length for the following reasons:

- data is stored in different data centers—including external storage;
- an increased number of handover points in the operation of a bank due to outsourcing and near offshoring;
- an increased information overhead in terms of metrics to be kept—for internal reporting, for customers, and for regulators;
- more information and a larger number of sources to be integrated.

Simultaneously, the latency tolerance on the availability of the information has gone down.

1.4   THE DATA MANAGEMENT PROBLEM

The current IT landscape resulted from an early start and successive waves of automation and technical innovation. Unfortunately, a lot of local work resulted in disparate formats, the use of different identifiers for instruments or clients, major challenges in extracting and combining information, and— last but not the least—the decentralized and often not clearly defined functional ownership of data. Typically, information is owned on a department

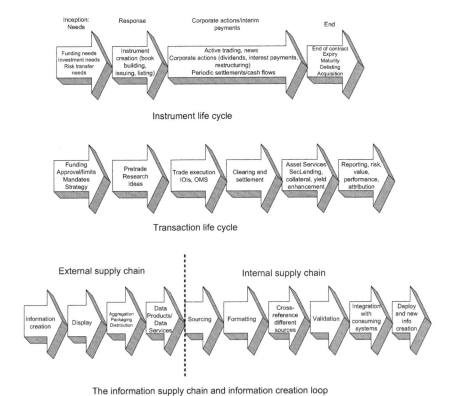

Instrument life cycle

Transaction life cycle

The information supply chain and information creation loop

FIGURE 1.5   The instrument and transaction life cycles and the information supply chain.

basis, so the equity department, the fixed income department, and the treasury department would all guard their own data stores even though there is overlap between the data they keep. Only recently has the importance of **data governance** been recognized.

The reasons for the typically convoluted architecture include:

- *Political reasons.* Institutions are often organized by client segment, by geography, and typically also by product line. Each outfit wants to control its own destiny and wants to decide where its budget on applications and content goes. Because time to market is often critical, the focus of front-office decision makers can be short-term oriented. Frequently, a solution is put in place with little consideration for the architectural implications or the cost of maintaining it in the future. It is typically easier to add a small application or content feed that you control than to be dependent on someone else.
- *Content licensing issues.* Many vendors of content cater for the office politics of their clients. No content vendor will want to give away everything at once so there is a tendency to license content at the lowest organizational

unit. A department has a problem and a budget. The budget can fix their particular problem and the issue is solved. The fact that there may be 10 other departments with similar problems is overlooked, or when it is recognized, the efforts to unite all these departments in the face of both internal (time to market) and external (enterprise-license-averse vendors) complications often prove insurmountable.

- The content industry lacks *standardization*. Vendors created their own formats and often their own identification and product classification codes. This complicates the integration problem. This cross-reference problem is exacerbated by the prevalence of local codes, different standards, and the reuse of some codes, which leads to ambiguity.
- *Embedded data models* in (legacy) applications. Applications embed their own proprietary data models and data stores, which makes integration very complex. All too often information, business logic, and presentation are welded together. The reason for this is straightforward: dependence on other parties to provide information adds a risk for the vendor.
- There are many *specific needs at department level*. In many cases pricing, processing, and modeling of complex financial products have become so specialized that niches in content and applications have been created. A diversified institution has a choice between buying multiple products and services externally and a very large internal IT and business analysis department. The job of integrating different products and content services is often left to the IT department or system integrator. Whipping data into shape, molding and massaging it to fit into often arcane formats and welding it to heterogeneous architectures, and subsequently re-interpreting and repairing it has taken up the bulk of overall IT budgets.
- Many institutions have *heterogeneous architectures*. One reason for this is the time-to-market pressures that cause institutions to buy applications that do not run on platforms they already have. Although there are integration products that address this situation, this has led to increased cost and a higher hurdle for integration.
- *Subject matter expertise.* The knowledge about specific content and business logic is concentrated in the hands of the users. As financial products have diversified along the dimensions of credit, interest rate, equity market, currency, and commodity risk, it is very hard to set up a centralized department with the required budget and staff to service these various users.

Typically, additions to the IT and content infrastructure of a firm lead to locked-in costs that are very hard to get rid of. IT budgets will tend to go up since the vast bulk of the budget is spent on maintaining yesterday's automated processes. However, this also means that the only new developments that can be funded are precisely the short-term tactical extensions that only add to the mess and do not structurally address it. Only massively disruptive internal or external events (such as the crisis and its new regulation) provide the opportunity to structurally address information management flaws. The discipline of enterprise data management (EDM) exists to do precisely that. Easier and faster access to information is one of the largest, mostly untapped productivity improvement opportunities left to financial institutions. Information becomes valuable only when it is acted upon and valuable pieces of information are often marooned in specific systems, users, or departments.

"EDM" refers to the ability to effectively source, verify, combine, and integrate data for internal applications, end users, and external reporting. It focuses on the creation of common master data sets that satisfy organizational quality criteria including accuracy, timeliness, and completeness. EDM aims to provide a common data foundation for a diverse set of internal and external stakeholders. It often encounters organizational difficulties as it cuts across different business silos and has to prioritize requirements and harmonize data definitions coming from different areas of a firm.

The goal of EDM is to achieve a common data foundation, create trust and confidence in data assets, and secure the integrity of the data used in business processes. Especially in financial services with its product proliferation and heavy requirements on internal and external reporting, improved EDM is a necessity.

Persons in charge of EDM initiatives or operations are typically called Chief Data Officers (CDOs). These strategies and persons have a number of challenges that confront them. The CDO often has one of two primary mandates:

The Evangelist CDO	The Policeman CDO
Small team	Budget holder, contender of COO/CIO
Soft powers	Line management
Reports to CIO or COO	Reports to CEO
Matrix organization	Centralized organization

The EDM and CDO's objective is primarily business enablement: improving the supply chain of information. This is achieved through a combination of improved information sets, toolsets, and processes. Knowing what you need where and when allows you to design and manage your processes in a better way. This process often starts by taking an inventory of data stores and data definitions to expose ambiguity—followed by a harmonization effort of definitions and technical standards. It can be followed by clarifying data ownership and change management procedures. Finally it is followed by the introduction of tools and techniques that help business enablement and that facilitate the distribution and controls on information—rather than locking down today's processes to block tomorrow's opportunity.

If that foundation is put in place, it opens the way to answer fundamental business questions in operations, and commercial and regulatory reporting more easily. Example questions are as follows:

● Will the client pay? How likely is it? What loan amount can the client service?
● Where to find new business?
● What else can I sell this client?
● What are my costs of servicing this client? Of keeping this product alive?
● Should the client trade this? Should I trade this? Are all the true risks to the business exposed?
● Are the products we trade set up correctly? Do the prices we have reflect the latest state of affairs? Did we process all changes to business entities, corporate actions, and instruments?

Cost pressures as well as competitive pressures on the industry make the ability to quickly and confidently respond to these questions by harnessing all available information more important than ever.

How big is the problem really and what does it cost the industry? The total spend on data is estimated at $26.5 billion (see https://burton-taylor.com/). The total IT spend for financial services is estimated at over $460 billion (2015 estimate at $461.4 billion for hardware, software, internal, and external IT services; expected to reach $534.7 billion by 2018 with compound annual growth rate of 4.7%; see https://www.idc.com/getdoc.jsp?containerId=FI251689). Estimates on the direct cost of poor quality data vary widely, partially due to difficulties and inconsistencies in defining quality: "quality variances ranging between 4 and 30 percent, including missing issues, missing data elements, inconsistent coding and mismatching data values" (EDM Council metrics; see www.edmcouncil.org). Annual data management costs for large firms have been estimated to be between $238 million and $1242 million (Source: Grody et al., 2006).

The overall challenge for IT departments in financial services is not that of decreasing settlement fails, cutting data cost, or complying with regulation. The fundamental challenge is of putting all the data they have to work and to empower end users.

A large proportion of information that firms source and keep is inaccessible to end users. It either is buried in isolated formats, databases, and applications or is unstructured information that flows in and out of firms without clearly impacting any process because it cannot yet be processed by the current infrastructure. Either way, the value of this data is unknown and is called **dark data**. It is data that is often hidden from end users and that could be obsolete, trivial, or valuable. Estimates suggest that around half of all data in a firm is dark data (see Global Databerg Report, https://www.veritas.com/product/information-governance/global-databerg).

The opportunity cost and the competitive impact of **not** using a significant portion of available data is huge.

1.5   OUTLINE OF THIS BOOK'S CHAPTERS

The book is structured as follows. Chapters 2 and 3 provide an overview of the different categories of financial data itself (Chapter 2) and the business processes that this data feeds into (Chapter 3). To do this it uses different perspectives including the trade life cycle, customer interaction, and regulatory reporting viewpoint. Chapter 4 provides an overview of the challenges and pressures on financial services firms' information management infrastructures including changing business, client, and regulatory demands against the opportunities provided by new technological developments. This chapter is about the "what needs to get done."

Chapters 5–7 discuss the "how": how firms can respond to these challenges from a technological (Chapter 5), data management process and quality management (Chapter 6), and organizational (Chapter 7) perspective. For example, cloud, data retention, and information security management will be addressed in Chapter 5 and the measurement of uncertainty of information and gaps in the record will be covered in Chapter 6. Management issues on information quality metrics will be addressed in Chapter 7. Important areas, such as data quality measurement and data governance will come back throughout these chapters viewed from these different perspectives. Chapter 8 concludes and provides an outlook on the future.

As an overriding theme for the discussion we take the *supply chain* view of financial information—that is to say the process, tools, and techniques through which financial institutions prepare the data gathering from a range of internal and external sources for usage by end users and reporting. This supply chain approach provides a binding theme for all the chapters and offers a good framework to discuss the *foundational nature* of information management. It is particularly relevant in the field of big data where specific data sets and metrics need to be distilled out of vast sets of unstructured data and where data quality management and the choice of the right summary metrics to manage against is important. While there is much more information on clients and markets available, the demands that customers and regulators put on preparation turnaround time, quality, and availability have increased correspondingly.

Many industry groups clamor for and promote data standard and common data models, instruments, and venue taxonomies. Yet inertia often remains since existing formats and systems are incredible sticky and costly to replace. Also the "if it ain't broke don't fix it" mentality of staff in operations still prevails. But this is not so much the question. It may not be "broke" but definitely the processes are leaky and erode profit margins. More importantly, they act as a brake on innovation and that opportunity cost dwarfs any operational cost of keeping suboptimal processes alive.

In the case of implementing data standards, there is an understandable reluctance to be the first. The first industry player to implement a new standard will hardly benefit. He or she bears all the risks and costs of implementing first, yet has *no one* to talk to using this standard, although there may be benefits internally of standardizing onto a standard into which a lot of thought has gone. Regulators are becoming less reluctant to prescribe specific standards as they themselves face their own information processing challenges.

At the end of the day, EDM should understand the requirements and service of the different parts of the transaction and master data life cycle. It should accomplish a full understanding of the content needs and of the information supply chain for the various stages in the instrument and transaction life cycle. Adequate metrics and reports should give insight into the state of the business. Questions, such as "Where are we? What are we doing? Where can we go wrong?" should be easily answered. EDM should help the information supply chain and data curation processes to hold up under regulatory scrutiny. Most importantly, it should put all of a firm's data to work, harvest any data that was previously "left out," resolve blockers, such as inconsistencies and definitional confusion, and provide a foundation for sustainable business growth.

REFERENCES

Grody, A.D., Harmantzis, F.C., Kaple, G.K., 2006. Operational risk & reference data exploring: exploring costs, capital requirements and risk mitigation. J. Oper. Risk 1 (3).

United Nations Environment Programme, 2015. Effects of financial system size and structure on the real economy. What do we know and what do we not know? Inquiry Working Paper 15/10.

Chapter 2

Taxonomy of Financial Data

Chapter Outline

2.1 INTRODUCTION

In this chapter we discuss different ways to look at financial information by function, by source, and by information type. We start with a high-level functional data model and then discuss the data creation process, the role of data providers, and the commercial aspects of the financial data business. The main part of the chapter is devoted to a clarification of the different types of financial information.

In the data model in the subsequent text we categorize different types of financial information, seen from the perspective of the transaction life cycle. We will go through each of these categories.

A Primer in Financial Data Management. http://dx.doi.org/10.1016/B978-0-12-809776-2.00002-8
Copyright © 2017 Elsevier Ltd. All rights reserved.

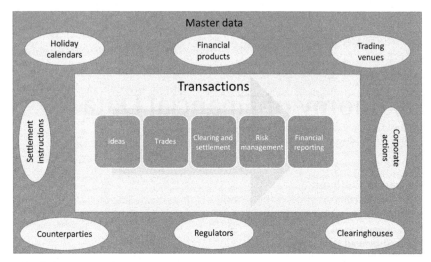

FIGURE 2.1 Master data and transactional data.

First, we need to distinguish between master data and transactional data (Fig. 2.1).

2.2 MASTER DATA VERSUS TRANSACTIONAL DATA

Master data consists of the necessary information sets to conduct business: the boundary conditions for transactions. It is the definitional data needed to set the stage before any business can be done. This includes the information to identify clients and information on the company itself. Second, it includes information on the type of business that can be transacted, that is, definitions of the products that can be sold to or bought from these clients. Third, it includes additional information governing the business that is done between the firm and its customers, for instance, opening hours, calendar information, and so on.

Transactional information is defined as any specific business between actors, between the nodes in the high-level data model mentioned earlier. This is anything that leads to a change in obligations between the two parties.

Master data is expected to change infrequently: clients may occasionally change their addresses and new clients may come in. The firm may sunset some products and introduce new products that can be offered. Transactional information is highly frequent information. It includes any money moved from or to an account. The number of transactions processed by a bank can run into the many millions per day. The systems designed to process and manage master data differ very much from those designed to process and manage transactional data.

In some cases there can be an overlap between transactional and master data. This is the case when two parties trade a unique instrument bilaterally, for instance, a complex OTC product. In this case the master data describing the instrument or contract is itself part of the transaction—the product traded does not live anywhere else apart from the specific contract between the two parties.

2.3 STRUCTURED DATA VERSUS UNSTRUCTURED DATA

A lot of change in financial data management comes from the introduction of unstructured data into decision making. Technological advances mean that data previously unavailable or out of reach can now be a part of automatic processing. Historically, information technology has been about organizing information into fixed structures: file formats, data dictionaries, and database tables. Increasingly, information is available for processing in unstructured formats. In Table 2.1 are present some examples.

Structured data has a preagreed format and set of domain values, for example:

- a date where the format can be for instance YYYYMMDD or DD-MMM-YYYY;
- a fixed set of ISO currency codes or country codes;
- a credit rating with a certain scale where each notch from Aaa to D has a specific interpretation;
- GPS coordinates;
- FpML or FIX messages for financial transactions;
- SWIFT messages for, for example, corporate action announcements or payments;
- a regulatory or financial report with a predefined structure, governed by a standard, such as XBRL.

Structures range from small scale (a set of three-letter currency codes, such as USD, GBP, EUR) to longer, such as SWIFT messages or financial statements.

Unstructured data can be anything that does not fit into a certain predefined template; this can include:

- tweets, text messages, chats;
- any free format text, such as news, a call report, a social media update, emails;
- pictures and video;
- audio, live conversation;
- sketches on a whiteboard.

TABLE 2.1 Structured Data and Unstructured Data

Structured data	Unstructured data
Fixed definitions and domains	Free format text
Database table definitions	Documents, news, emails
File formats and conventions	Video, audio, images

Note that there is often **some** underlying structure to unstructured data. For instance, a tweet has an upper bound of 140 characters and meeting minutes or emails do follow a certain social convention. Also, there is certainly a deeper **technical structure** in the subsequent text. Free format text may be stored, for instance, in ASCII character codes; audio may be stored in the mp3 format and a picture could be a .jpeg or a .bmp.

Sometimes we already have a blend: such as emails having fixed structures, or an email with an attachment XLS or a swift message composed of very specific fields but that also contains some fields of type "narrative," meaning free format text can be entered instead of a carefully controlled set of choices.

Structured or unstructured data does not mean simple or complex. Structured data has arisen, and has been imposed (even at the expense of precision and nuance) to enable automation. The boundary of data can be processed is shifting upstream thanks to big data technology, data can be mined earlier and earlier in its curation process. A major change recently has been that it has become far easier for business intelligence tools and reporting tools to get to work *directly* on the unstructured data. Recording, storage, and computing power have become cheaper, meaning that there is a lot more unstructured data available that can be harvested for use.

2.4 SOURCES OF FINANCIAL INFORMATION

2.4.1 Classification by Source Type

If we look at the overall financial content industry, we can distinguish between four main sources of information. A summary and examples of data categories of each of these four types of sources is provided in Fig. 2.2.

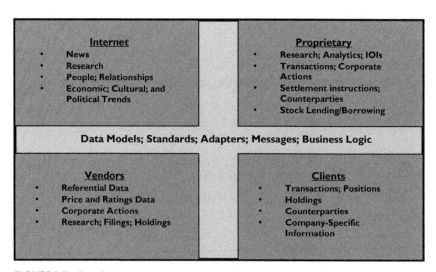

FIGURE 2.2 Data by source type.

This breakdown categorizes data in information supply chain terms. It can be obtained in one of 4 ways:

- public internet (News, Research, People; Relationships, Economic; Cultural; and Political Trends);
- vendors (Referential Data, Price and Ratings Data, Corporate Actions Research; Filings; Holdings);
- proprietary (Research; Analytics; IOIs, Transactions; Corporate Actions, Settlement Instructions; Counterparties, Stock Lending/Borrowing);
- clients (Transactions; Positions, Holdings; Counterparties, Company-Specific Information).

Each of these four categories plays a large role in providing the information that is needed to run the instrument and transaction lifecycle processes in a firm. Technological changes have increased the distribution options and have made it easier to create and distribute new content products. Increased access due to regulatory disclosures has caused some information categories to become more commoditized and has eased access to them. On the other hand, new financial market and product development has also led to many highly specialized data services.

2.4.1.1   Public Sources of Information

Public information includes all content freely available either because of regulatory disclosure requirements (e.g., filing, prospectuses) or because it concerns government published information (e.g., macroeconomic numbers). It includes chamber of commerce information and public filings on financials (cash flow statement, income statement, balance sheet). The rise of the internet and online databases has greatly facilitated the ease and speed of access to these materials. However, in many cases data vendors add sufficient value collating public information to make a marketable product out of it. This includes press releases, patent office information, information published by government or central organizations, such as macroeconomic statistics, or interest rate changes from the relevant central bank. It can also include informationthat influences the pricing of products: macroeconomic data, such as GDP numbers, weather data from government meteorological offices, or perhaps geographical data. Other relevant information includes rulings from courts or competent authorities, policies from regulators, verdicts in lawsuits, information on bankruptcy cases, and so on. Apart from this highly structured information where the exact coverage is often dictated by regulation or procedures there is another set of less reliable public information: a lot of this includes gossip, hearsay from forums and blogs.

2.4.1.2   Commercial Data Vendors

We can make a distinction here between *data providers* (which actually *create* content) and *data vendors* (which specialize in *aggregating and distributing* that data). The first category ranges from niche vendors that specialize in price evaluations of a certain class of derivatives to rating agencies that rate instrument issues only in a lesser developed country. The second category ranges from specialized players that collate and reformat highly focused information (e.g., MBS pool information from public sources, such as the US government–associated agencies) to the handful of large-scale, globally active aggregators that have hundreds of sources including rating agencies, brokers, exchanges, governments, and banks.

2.4.1.3 Proprietary Data

Every financial institution generates enormous amounts of proprietary information. This constitutes the firm's "edge," on which business decisions are based. It includes the body of specific and private knowledge on customers (such as specific credit information), knowledge on the pricing of specific instruments, the institution's strategic planning, the current state (the current portfolio), and its own research through proprietary financial models from macroeconomic developments to specific internal credit ratings. Apart from elements that provide the institution its competitive edge, proprietary information also typically comprises information that is needed to run business processes, such as Standing Settlement Instructions (SSIs), unique identification of counterparties, cross-references between different instrument identifications, identifiers for OTC instruments outside the current ISIN (ISIN stands for International Securities Identification Number, ISO standard 6166) scope, and so on. This kind of data, although proprietary, is generally not competitive, and could potentially be shared with peers to improve operational efficiency.

2.4.1.4 Client and Counterparty Sourced Content

The last major category is that of content that is sourced through the business relationships an institution has. First of all, this category includes data that defines the relationship, such as what products can a client trade, what is the composition of its portfolio, what is the benchmark against which investment returns are assessed and what execution strategy has been agreed. In addition, and in particular for the professional trading counterparties, this includes "master" agreements or legal frameworks that provide for the netting of transactions and other settlement-related information, such as SSIs. An interesting development in content sourcing has been the *pooling* of information between peer groups to arrive at a higher data quality and more comprehensive picture together at lower cost. Pooling can work for data categories that each institution would typically gather itself on a piecemeal basis and for which no off-the-shelf commercial content products exist and which constitute data sets that are not the basis of competition. Examples include:

- nonsensitive information on legal entities (such as location, legal structure, address details, identification, country of incorporation, and full legal name from the articles of incorporation);
- settlement instructions, such as location and account information, payment instructions, and perhaps standing instructions on corporate action handling;
- information on operational risk, such as losses. Loss data is required for all banks that want to do their own modeling on operational risk and is hard to get. On the other hand, a large set is needed to arrive at statistically meaningful estimates.

In the last case, as long as the data set is anonimized, pooling can help. For the customer data as well, pooling can reinforce quality and lead to a more comprehensive view. Note that competitive information on customers, such as credit assessment (internal ratings) would be less suitable to share as this is closer to the core business of a bank. Advantages of peer-to-peer sourcing are that it is cheaper and it will also be reliable as it comes from people who are using it on a daily basis.

2.4.2   Short Overview of Data Vendors

Fresh content has been compared to the oxygen supply of the financial markets and is a necessary condition to any decision making, whether it means doing a trade, informing an investor, selling a product, or complying with regulation around the instrument and transaction life cycles.

Many companies have made it their core business to either create content or package and distribute it. In the first category we find, for instance:

- the rating agencies that provide information on instrument and company level;
- research groups that look at the outlook for certain industry sectors or regions;
- index providers that construct indices that reflect certain markets or that speak to specific investment strategies.

The second category of packagers contains many well-known names in the financial industry, such as IDC, Thomson Reuters, Six Financial Information, and Bloomberg. Companies, such as Thomson Reuters have large data collection departments to gather data from hundreds of different trading venues, exchanges, brokers, and so on. Newer content players can set themselves apart by becoming market leaders in certain subsegments. Estimates on the total revenue in the financial market data, analysis and news space go up to over $26 billion (see https://burton-taylor.com/ for 2016 Financial Market Data/Analysis Global Share and Segment Sizing report).

2.4.3 Fragmentation and Consolidation of Information Sources

Between creators of data (market makers, research agencies, exchanges, and issuing parties) and end users, there is a complex and often lengthy information supply chain. Information becomes useful only when it is acted upon or used to create new information. Knowledge about the information supply chain, and about changes and additions to information at each step is a condition of success for any activity that critically depends on this information. Unfortunately, all too often information arrives too late or is incomplete or degraded at the place where it is needed. Information even can get lost on the way and get stuck at a certain stage in the supply chain—for example, when incomplete records are passed through or when incorrect assumptions are made.

The transaction life cycle depends on internally and externally sourced information, and the instrument life cycle often on external data. The external part of this chain includes the ultimate sources and content intermediaries, such as information aggregators, validators, enrichers, and distributors. An institution often sources content from many different providers and has to do substantial internal integration and quality assessment work. After purchasing, integrating, and cross-comparing, the information supply chain continues internally until it reaches users or application where it can be acted upon.

Effective content management is about getting the right content to the right place at the right time—for the lowest possible cost. The more intermediaries in the supply chain, the larger the time delay and the higher the potential for error.

The content market will continue to grow for various reasons:

- Technology advances make that more data can stream into an application or pricing model within a certain time frame.
- Pricing models and risk models are increasingly data hungry and operate on more granular data.
- New financial product development requires more prices.
- More data is pushed into the public domain—including transactional data from OTC products.
- There is more emphasis on accurate and timely reporting from both regulators and customers.
- A proliferation of execution venues due to new venues and internalization by banks leads to new data sources.

As financial services firms are consolidating and globalizing, the structure of the content industry is changing as well. One example of this is a smaller amount of country-specific products but a move to more global information products, both functionally in terms of content and also in information providers' sales and marketing strategies.

However, the competitive dimensions for content providers remain largely the same:

- *Coverage.* What financial products, what markets, and what other data types are made available through one channel and one data format? To what extent is a one-stop shop—commercially and technically—provided?
- *Unique content.* What content cannot be found anywhere else? Specific indices, credit research, and so on.
- *Accessibility.* How easy is it to integrate the data into processing? Can one or more steps of the often lengthy supply chain be skipped? What access options are available, for example, APIs and services, to retrieve content dynamically? Can data retrieval be tailored to the institution's needs, for example, on a portfolio basis? Can subselections of attributes be made to limit unnecessary data flow and spend? Are data standards used?
- *Value add.* Is sufficient value added in cross-referencing and linking between different data elements offered? Often insights and value can be created by providing cross-reference and structure over multiple types of data.
- *Latency.* How timely are products delivered? Timely can refer both to timely batch delivery so that information arrives in time to be used in a daily batch process and to low latency for streaming data and the maximum time lag in relaying price-moving information, such as press releases, earning, and corporate actions.
- *Quality, pricing, and service.* How reliable is the vendor? What is the quality of the information? Do the quality dimensions meet the institution's needs? What is the overall pricing and service (this also includes training and support) around the product?

2.5 DATA CREATION PROCESSES AND INFORMATION LIFE CYCLE

2.5.1 Data Origination

The total number of information sellers runs in the thousands and includes many sellers of local data (exchanges) and specialized product data (research firms). The number of large aggregators that provide comprehensive content solutions is very small. There are large barriers to entry and to become a major all-round packager and distributor of data requires economies of scale. New entrants in this space need to have an edge either in cheaper and faster aggregation of information or through offering it via new channels and in new—technical or commercial—models.

Data is being continuously created and new financial information can arise because:

- New products are created: companies list, governments and corporates issue debt, and derivatives exchanges create new sets of futures and options per a predefined calendar.
- New clients are onboarded; new companies are founded.

- Corporate actions: companies get bought, stocks splits, and companies go bankrupt and redeem bonds early.
- News and research reports are published, financial statements are published, people change jobs and firms merge or spin-off entities.

One important trend enabled by information technology is that of disintermediation. Whereas content creators used to rely on aggregators to reach their users, now more players enter the market to sell data directly. This includes firms that already had large amounts of data collected as a by-product of other activities, such as clearing and trading but want to realize some or more of its value.

Large Exchanges, Brokers, Central Securities Depositories, and Custodians are all developing data businesses alongside their traditional businesses. Exchanges and brokers were earliest in distributing their real-time pricing information. Other examples include:

- content offered for free as a part of marketing strategy (custom indices provided by some of the banks to signal market leadership in a certain product set);
- data offerings combined with technology offerings, for example, integration of content with a trading, risk, or portfolio system.

In some cases regulation is fostering the creation of new data businesses, for instance, through pushing new content into the market. This includes, for example, trade repository information and new venues created as a result of MiFID II (MiFID II is new European regulation aimed at improving the efficiency and transparency of European financial markets and increasing investor protection).

Some banks have also started to sell not just data but also value-added services in tooling and analytics on top of it. This can be derived data sets, such as risk factors or evaluated prices coming out of their own quantitative models but it can also include tailored risk advisory services based on specific portfolios. On the other hand, pure information providers also move to downstream vertical integration, for example, by expanding into data display terminals and applications.

2.5.2 Data Vendors and Licensing

2.5.2.1 Licensing and Contract Management

License agreements are a way for content providers to protect revenue and intellectual property, and to secure future upside. They are often structured to correlate to the perceived value by the client.

The market for content is large and growing. The length and complexity of the supply chain means that content can enter an organization at different points in the processing. Some content will be directly distributed to a desktop but other content will be processed automatically and can be distributed enterprise-wide. Given the size of the content market earlier, appropriate licensing and digital rights management is important. In this section we discuss licensing issues around content and various pricing models.

2.5.2.2   Entitlements and Contract Databases

Entitlements refer to the administering of rights on content and the reporting on data usage. Since many data owners, such as exchanges have started to appreciate the value of the content and are becoming more assertive about monetizing the value of their information products and are demanding more reports on data usage by user or by application, it is important that users properly control and report on the usage levels of the data. Products, such as Thomson Reuters DACS (stands for "Data Access Control System"; see https://customers.reuters.com/developer/rmdsandtools/dacs.aspx?) can keep track of data use by different applications. The advent of digital rights management will secure a more prominent place for such applications in the future.

Data vendors do not always have control on the downstream distribution of their content products as users can copy and paste information and rout it throughout the institution or even outside. Contracts are not always clear and unambiguous and there tend to be shades of gray. We can distinguish between different types of "revenue leakage" for content providers (Source: FISD; see http://www.fisd.net/presentations/20031205redistributionoutline.pdf):

- There can be uncontrolled internal redistribution of data. Data may be passed on to users and applications outside the contractually stated use.
- There can be unreported external redistribution. For example, an institution can use price data in balance and portfolio reports to external clients.
- Data can be scraped or shredded from display products, such as web pages, terminals, or broker pages. This practice, when not authorized, is also called *data snooping*.

Apart from this, there is also a fine line between commercially sourced information and any derivative content created by a financial institution's own proprietary models and content for which the vendor's content served as one of the inputs. When price data is used as part of the input to create something else, for example, an index, or to price a derivative based on proprietary models, it is not always clear at what point a vendor's intellectual property claim ends. Vendors could claim that their content is critical whenever the result would have been different without their input. In any case, market data use is more and more carefully audited especially in times of low revenue growth.

2.5.2.3   Data Ownership

Just as sometimes institutions may have overused certain data, they have also frequently paid for data they did not use. With procurement and sourcing becoming more and more centralized, these mismatches have becomeclearer. Another reason that has brought these contractually permitted and real usage discrepancies to light has been the decrease in the number of processing centers of an institution as IT is consolidated and redundancies in processing addressed. Institutions and their data users have to carefully match the usage needs of a department to available data services in the market. Contracts and the invoices to which they lead can be very complex. Therefore, contract and spend management for the content industry is a niche industry in its own right. Contract management application providers and consultants in this space offer consulting services on (market) data needs and packages, such as contract databases that include information on commercial data products and in which a financial institution can keep track of its own purchased data inventories [examples include The Roberts Group (https://www.trgrp.com/), MDSL (https://www.mdsl. com/market-data-management), and Screen consulting (http://www.screenconsultants.com/home.htm)].

2.5.2.4   Licensing Models

All pricing models are about striking a balance between serving the customer properly and getting value out of an account for the supplier. They also serve to preserve potential upside for the party that is selling something: vendors normally do not want to sign up to an enterprise-wide, unrestricted, perpetual license to all the content they can offer because that effectively takes away an account from the market for the future. However, attractive such a deal may look short term, invariably it is a bad idea long term.

We can distinguish between the following licensing models:

- *Itemized Pricing.* This refers to pricing commensurate with the number of items and how often they are accessed. Often this is the case in "portfolio"-type content products, that is, products that deliver information against a customer's particular universe by a list of CUSIP, ISIN, or SEDOL codes of instruments. Pricing then often is a dollar amount per item per time window, for example, a month. A variation on this model is a pay per use or view report whenever information is requested or an API is hit. We find this category at vendors that used to supply hard copy reports, for example, on credit score or on the legal structure of a company.
- *File-Based Pricing.* A less granular approach compared to itemized pricing is pricing on a per file basis, which can be done when a data product is partitioned into different categories. A product can, for example, be broken up into chunks representing different markets, geographical areas, or asset classes.
- *User count restrictions.* In the case of usage restrictions, pricing is not so much done on content basis, but on a user count basis. There could be differentiation between *display and nondisplay pricing*. Especially in the area

of real-time data, firms distinguish between charging for display use and automatic processing, which means capturing specific charges on "eyeballs." Content can also be licensed by department, by number of named or concurrent users, by sites, and by countries. In a user count model, the unit of measurement has to be clearly defined. It could be a generic user account or a natural person and there could be cases of persons sharing the same password/user ID. To what extent a vendor restricts this depends on the product. It does not make a lot of sense for a targeted niche product, but does make sense for the larger aggregators.

- *Use case based.* Tied to business processes or specific reports to regulators or clients. Especially when the content is used in external reporting to clients.

As there can be multiple intermediaries between the source of the data and the end user, responsibility for the unit of count of usage and for reporting it back to the source needs to be clearly defined. Typically, the final intermediary or end user reports back to the originator.

A vendor's flexibility in pricing schemes increases with the size of the account and the largest accounts will be able to negotiate package deals that go beyond content delivery to include services, such as training, support, or customized content products. Discounts for global licenses or site licenses are typically available; on the other hand in more difficult times opaque licensing agreements provide content providers with several angles to preserve commercial upside by limiting license models by consuming systems or use cases.

Some of the key characteristics of pricing models are listed in Table 2.2.

The pricing model should first of all distinguish between systematic consumption of market data (through applications) and individual consumption (through terminals) and will also be related to the way the content is made accessible.

TABLE 2.2 Characteristics of Different Content Pricing Models

Itemized pricing	Bulk pricing	User count restrictions	Use case based
Physical delivery, for example, reports. Per instrument, portfolio. Tied to a service model. Attractive when starting new services as it scales with usage	Per category, per market, per asset class. On a geographical or sites basis (enterprise wide is the extreme case)	Pricing tied to physical desks, specific user names, number of concurrent users, and/or licenses that limit usage to consuming systems or departments	Licensing tied to activities, for instance, valuation, risk, securities lending, client reporting, regulatory reporting

2.5.2.5 Data Carriers

When content is made available through a file, the vendor can lose control over where it ends up as it can be endlessly copied and sent throughout an organization. Enterprise-wide agreements can be struck to prevent this problem. Consuming companies will probably still need to count usage to allocate costs internally.

If content is made available through a password-controlled display facility, a vendor can exercise more control. In *browser-based access*, instead of physically delivering the content through a file or through messages, the content can be looked up through a web browser. This way, users can look up information when they need it. This can be useful to look up reference information. A related way of *user-oriented pricing* is providing information via *terminals*. In this case, access to data is tied to specific desks. Vendors do not always want to police their users, so controls need to be in place to prevent underreporting on data usage in terms of number of terminals.

Terminal products typically combine content with applications, for instance, for portfolio management, research, or trading. Advantages of these distribution methods are the extensive customization of querying that is possible and the potential combination with analytics and other applications. Here, information is directly brought to the end user but cannot be easily used in an automated process. The pricing chosen for file-based products will typically reflect anticipated wider usage.

Some vendors publish different versions of the same content, typically in different quality grades. A common example is that of immediate versus delayed streaming exchange data but content products can also be offered in a full detail or summary version.

2.6 OVERVIEW OF INFORMATION SETS

In this section we will survey the various content types listed in the data model with which we started this chapter. News drives prices and triggers transactions. Legal entities have relations with other legal entities, publish financial statements, have credit profiles, and have traders or portfolio managers who run sets of positions or portfolios. Communication logs of traders are kept for compliance reasons. Portfolios are updated by transactions and trade support data is required to execute transactions. Portfolios are sets of holdings in financial instruments on which various analytical risk, liquidity, and performance measures can be defined. Tax information can apply to positions, transactions, and instruments themselves (Fig. 2.3).

2.6.1 Security Master Data

Security master describes the terms and conditions of financial products. This can range from relatively simple (for FX and deposits) to a large number of terms and conditions (for fixed income and structured products).

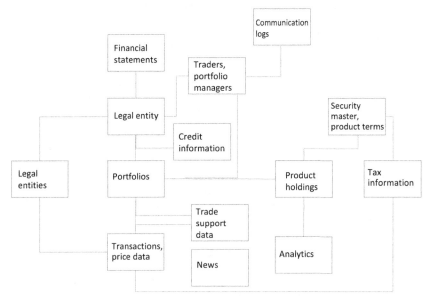

FIGURE 2.3   Overview of financial information types.

Financial products (Fig. 2.4) are created to:

- bridge the needs of corporates or governments needing money and investors willing to part with their money for some time—for example, stocks and bonds but also deposits;
- provide exposure to certain risks for hedging purposes—for example, options and futures;
- provide direct access to a financial commodity needed—for instance, FX;

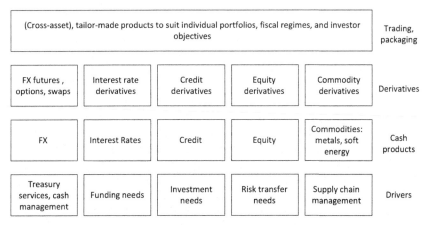

FIGURE 2.4   **Financial product families.**

- provide financing optimization by swapping different interest rates—for instance, swaps and swaptions;
- provide fee income to the providers of these products;
- take advantage of provisions in the tax code, such as subsidies on "green" investments;
- a combination of the above.

There are different ways in which financial products are classified. The most common classification schemes include:

- Classification of Financial Instruments (CFI) code or ISO standard 10962. This is a categorization at multiple levels that goes from high level (Equity, Debt, Futures, Funds, Index, Miscellaneous) to very detailed using a six-level hierarchy. For example, ESNTPB stands for Equities/Shares/Nonvoting/Restrictions/Partly paid/Bearer and OPASPS stands for Options/Put/American/Stock/Physical/Standard.
- ISDA has published taxonomies for credit, interest rate, commodity, and foreign exchange derivatives (see http://www2.isda.org/functional-areas/technology-infrastructure/data-and-reporting/identifiers/upi-and-taxonomies/).

New instruments are created from scratch, such as new bonds or equities, are periodically created according to predefined rules and timetables (exchange traded derivatives), or are created ad hoc between two parties in the form of a bilateral contract. Common fields for the main CFI instrument types include:

CFI category	Common fields
Equity	Dividend, nominal amount
Bonds	Maturity date, coupon rate, currency
Fund	Main holdings, fund manager
Future	Expiry date, underlying
Option	Call/put, exercise price, expiry date, underlying
Miscellaneous	Swap currency, swap tenor

Changing needs of investors and changes in tax codes mean that there will always be new parameters that can describe a financial instrument.

Many data vendors that offer larger aggregated products around security master have data dictionaries that contain thousands of different fields. Especially in structured products, large numbers of parameters can be needed to fully and accurately describe the structure. An issue in itself is the identification of financial instruments. For securities the most commonly used format is the ISIN code. OTC derivatives are by definition bilateral contracts and not separate products so here identification is difficult. This is being addressed by a harmonization attempt to come to a standard Unique Product Identifier [run by the Committee on Payments and Market Infrastructures (CPMI) and the International

Organization of Securities Commissions (IOSCO)] Also, MiFID II expands the scope of the ISIN code to OTC products.

Index information is a separate category. Indices provide summary views of a certain broad market (FTSE100, S&P500), region (S&P Latin America index), or an industry sector (MSCI World Banks Index). Indices are not only useful as a proxy for a broader market but also often mirrored by investment products. In that case we call the index "investable." The rise of the ETF market has given index providers a big boost and is an important source of revenue. Indeed, indices are not only market statistics but also very powerful brands.

2.6.2 Trade Support Data

A second category of financial information can be summarized as "trade support data." With this we mean information needed to facilitate the trading and settlement process that includes SSIs (various services including Alert from OMGEO pool this information from the institutions to make it available to broker/dealers simplifying this data collection task; see www.omgeo.com), master agreement information, holiday calendars, market codes, and company/person directories.

Standing instructions refer to a client's instructions regarding not only where and how to settle (which accounts, which currencies) but also what to do with dividends and other capital distributions, how to handle excess cash and the proceeds of sales of securities, proxy voting at shareholder meetings, and default choices if there is the option to choose between a cash and a stock dividend.

Holiday calendar data contains the business day conventions of a country or region, for example, listing the typical working days, for example, Monday to Friday, or Sunday to Thursday, and the bank holidays when trades cannot be settled and can also include trading hours of exchanges. Holiday calendars can be organized by a geographical indication, such as country, region, or city, or by trading center, such as the exchange. Holiday information is often more complicated than thought at first because:

- Holidays can be dependent on the occurrence of other holidays during the week or in the weekend.
- Holidays are sometimes announced at short notice, for example, to coincide with an election or other political event.
- Some countries have a mix of holidays from various religious calendars.
- The authority that determines the holiday varies. Sometimes it is enshrined in law; sometimes it is the central bank, a religious authority, or the government.

Holiday information is needed not only to know when to settle but also to calculate the number of business days [major providers of holiday information include Swaps Monitor (http://www.financialcalendar.com/) and Copp Clark (http://www.coppclark.com/)] for interest rate calculations.

Markets are identified using the Market Identification Codes (MIC) standard, ISO standard 10383 which lists all regulated markets. This is also often used in standing instructions to specify the list of eligible markets that a client wants to use to execute a certain order, such as the buying of shares.

Master agreements specify the legal framework in which two parties trade. These agreements specify how positions are valued, which products can be traded, under which conditions collateral needs to be posted, and how disputes are settled. The most commonly used agreement is the ISDA master agreement (see http://www.isda.org/publications/isdamasteragrmnt.aspx).

Trade support data also includes operational information to help with post-trade processes including:

- Information on SSIs of firms with account information. The counterparty of the trade is also part of the required settlement information.
- Eligibility of clearing. Which securities are accepted by clearing firms, such as Clearstream, Euroclear, and SIS?

Settlement and payment instructions indicate how to rout payments or where to settle certain products for certain counterparties. This category of information is typically pooled from the market players who have to specify where and how they want their trades settled. Settlement data includes the following subcategories:

- Commercial payment data. This includes the SWIFT BIC directory (bank identification codes) and includes the CLS bank directory for settlement of foreign exchange trades.
- On securities' settlement instructions, there are several poolers of account information, such as Omgeo (see, e.g., http://www.omgeo.com/solutions/product.php?s=22&p=HJROUEAUGZ_1_0).

Many institutions are only now starting to consolidate their settlement information. Historically, changes in settlement instructions were communicated over the phone, by fax, or by email. The purpose of electronically sourcing settlement instructions through aggregators is to increase straight through processing (STP) rates, transactions processed without any manual intervention, and rates to cut costs. Any failed transaction will lead to costs of staff that has to investigate and possibly claims from the counterparty, such as interest rate charges for late payments (services to improve operational efficiency in settlement are offered, e.g., by SWIFT; see https://swiftref.swift.com/ssi-plus).

2.6.3 Corporate Actions

Corporate actions data is information about changes in products or companies. They include:

- Expected income events. For instance, the payment of a quarterly dividend on a stock or the payment of a coupon on a bond. The amount of the dividend will be announced beforehand; the amount of the coupon either is already defined in the bond's legal documentation or, in the case of a floating rate note, will depend on prevailing market rates.
- Unpredictable income events. These are special dividends outside of the normal dividend calendar for a stock or, in the case of a bond, early redemption. A bond may be callable and redeemed earlier.
- Events at corporate level have to do with change of control or a refinancing of the business. When a company issues additional equity, this could be in the form of rights issues for stock, giving existing shareholders the right to buy newly issued shares at a discount. In a merger or acquisition, shares could be swapped for other shares.

Although there are common message formats, such as ISO 15022 and ISO 20022, the need for data cleansing is especially strong in corporate actions. The reason is that M&A processes can be complex and this is reflected in corporate actions. Moreover, quite often there are **choice** events, meaning the holder of the bond or stock has to respond to a corporate action. This could be, for instance:

- choosing between a cash dividend and a stock dividend;
- choosing whether to agree to a tender offer;
- making use of put option clauses in a bond to call for early redemption.

The timing is important since these choice of events come with a deadline.

The increase in volumes and complexity of events has meant a requirement for specialist labor, for "people with data in their blood." In choice events (tenders, choice dividends) there is a dialogue between the bank, the custodian, and the owner of the securities; therefore the information supply chain gets much more complex. The total annual number of corporate actions is estimated to be around 4 million; 3 million of these are scheduled interest payments and maturity events (for information on the number of corporate actions see for example the Oxera report on http://www.oxera.com/Oxera/media/Oxera/downloads/reports/Corporate-action-processing.pdf?ext=.pdf). Of the remaining 1 million, the bulk are dividend announcements. What is left are the tricky cases, such as the roughly 500 rights issues a year that normally require manual intervention. The more intermediaries, the greater the risk of information loss on the way.

One missed action affects a large number of positions, so the consequences can be dire. Especially missed voluntary actions can have very costly ramifications. Custodians typically source directly from local market sources that are close to the issuers and their agents. Front office needs corporate actions as one of the price drivers of securities. Investment managers need the best forecast of the dividend yield. Time series of financial instrument prices need to be corrected for capital distributions and splits for risk measurement purposes.

The different perspectives depending on the role that financial institutions and departments have incorporate actions are also reflected in the content products. There are pure corporate actions feeds as well as larger content products that contain corporate action information as ancillary to the security master. Some custodians or very large investors have been able to offer integrated corporate action products in this space that compete with the pure aggregators. They can use the information they already need to run their own business.

2.6.4 Prices, Quotes, and Liquidity

Pricing data comes in various categories, which we can distinguish by their:

- Delivery frequency and delay. This can include real-time data streaming straight from the point of origination, snapshots, end of day/close data, and historical data.
- Price type. A reported price can be an indicative quote, a firm or committed quote, a price at which a transaction took place, a price delivered by a model, a price that is a consensus based on a poll of multiple contributors, or a close price calculated according to certain rules (e.g., by a stock exchange).

We now discuss the various price data categories in more detail:

Real-time, also called streaming, data refers to a direct quote and price stream from a broker or execution venue. It will typically include not only prices but also time and sales information detailing the time and volume of transactions and order book information. The depth of the order book that is disclosed depends on the exchange. The regulatory trend is to make this pretrade information more transparent (especially with the upcoming MiFID II/MiFIR in the European Union that requires trading venues to post their pretrade information; financial services firms that hold a market share in a product above a market share threshold are classified as Systematic Internalizers and will be required to quote two-way markets). The driver behind this is investor protection: to make sure the investor has the optimal means available of price discovery to get the best available price. In case an instrument trades on various execution venues, an institution needs to consolidate different quote streams and order books internally to get the full liquidity picture.

Other dynamic pieces of information closely related to pricing are volume and open interest. Open interest refers to the total amount of open options or futures contracts. The volume is important to judge potential market impact of large orders. Streaming data is often made available in real time at a premium price and distributed for free or at a greatly reduced price in a delayed mode. The delay varies by exchange but can range from 15 min to several hours.

Close price. Regulated trading venues, such as exchanges report official close prices at the end of the trading sessions. This is done in a transparent way based on the last moments of trading activity—according to a procedure specified beforehand in the rules of the exchange. There are many different methodologies in place that various exchanges use to arrive at a close price. A close price could be defined, for example, as the VWAP of the last 5 min of the trading session. Sometimes a close price is simply defined as the price of the last trade or the best available price at the time of the last trade, for example, in the case of NASDAQ. Market indices calculated by index providers, such as Dow Jones and MSCI also have an official close.

Snapshots. For OTC markets there is no official close price; therefore the institution that has to revalue a position has to take a snapshot at a certain cutoff point. Consistency in time for all instruments in the portfolio is important. Sometimes an institution takes streaming data from an interval around the desired cutoff point and takes, for example, the median or the average price to prevent outliers at the snap time from distorting the picture. Sometimes an institution takes, for example, the 16.00 price from a respected market maker in that particular instrument. Sometimes a combination of both is used, for example, take the median price of a set of trusted market makers approved by risk control.

Historical data. Historical end-of-day data can be used for risk management purposes and also to calculate the historical volatility needed to price financial products. Risk management rules often rely on the use of anywhere between 1 and 10 years' worth of historical data. Similarly, historical data is also used for stress testing by applying the largest shock found in the historical record to calibrate the required regulatory capital. Historical data includes end-of-day data of financial products, but also historical tick data for market impact analysis and histories of other risk factors, such as macroeconomic variables, arrears rates, and default rates. Specific historical episodes that play a role in stress-testing scenarios include the 2008 Lehman bankruptcy, the European debt crisis, and earlier scenarios, such as the 1998 downfall of Long-Term Capital Management, the 1997 Asian Crisis, and the 2000 dot com meltdown and the Lehman shock in 2008. Regulators have moved to prescribing specific scenarios—either based on historical events or more exploratory [Bank of England, Federal Reserve CCAR/Dodd–Frank Act Stress Tests (DFAST), and European Banking Authority (EBA) program]. Apart from using it to see the effect of shocks on the portfolio, historical price data can also be used to backtest portfolio and trading strategies.

Historical intraday and tick data can be used also for the backtesting of strategies that rely on the microstructure of the market and can be used to study market impact and to measure liquidity. In addition to this, historical data is used in the pretrade area for research and for technical analysis. (To what extent the past is relevant is almost a philosophical question and a matter of taste. It is about data vs. a judgmental valuation.)

Price histories of some of the major exchanges go back a long time. In the case of the Dow Jones Industrial Average, the available history of stock indices goes back to the 1890s (only General Electric is left of the original 30 constituents of the index; stock baskets of other major exchanges have seen a similar attrition rate). The history of credit ratings goes back to the early 20th century when John Moody realized there was a market for independent company evaluation and started rating railway bonds in the United States. This can be seen as early outsourcing of financial statement and credit risk analysis. (Longer time series than that of the DJIA exist in economics as well as in meteorology. A graph charting Dutch government bond yields has been compiled by Rabobank and goes back to the 1580s. Time series in weather data started in the early 1700s when accurate temperature observations started to be recorded daily in both the Netherlands and Potsdam.)

When a portfolio needs to be revalued, some assets are easier to value than others. In case of illiquid instruments, "mark to model" is sometimes the only option. Normally, banks employ one of two approaches. Either the quotes come from an independent middle office function that collects them from trustworthy sources and signs off on any models that may be used or the quotes to revalue them come from the traders and the middle office compares these against, for example, a neutral market source.

Altogether, there are at least five different approaches to value the assets on a portfolio:

- *Evaluated prices.* This represent an opinion as to the accurate price that can be based on the instrument characteristics, market conditions, and the price of comparable products at the time the evaluation is made.
- *Contributed prices.* Where a set of contributors provides input to arrive at a composite price. A pool of contributors can lead to an average price, a median price, or the calculation of the average after the highest and the lowest price points have been discarded.
- *Trade prices.* Prices from transactions that the dealer has done are used; the trader reprices his or her own portfolio. There is a regulatory trend toward using "real" prices when it comes to risk and valuation. Simultaneously, more "real" prices are available due to regulatory initiatives in OTC market transparency. (This has led to new data repositories in the United States; see http://www.cftc.gov/industryoversight/datareposito-ries/index.htm and trade repositories in Europe as well that came out of clearing firms; see, e.g., https://www.regis-tr.com/ and http://www.lseg. com/post-trade-services/matching-and-reconciliation/unavista/unavista-solutions/unavista-trade-repository.)
- *Theoretical price* or *fair value.* This refers to an "objective" valuation of the cash flows discounting them using some discounting rate and credit assessment.
- *Exchange data.* Use official close data from a regulated trading venue.

In the pricing process, the breadth of coverage, timeliness, and accuracy are important criteria to evaluate data services. When a portfolio is revalued, the prices also need to be consistent. Consistency refers both to timing and to taking the same side of the quote. Picking the "Ask" quote when you are long and "Bid" when you are short would misrepresent the value and be too aggressive. Also the time of the quotes needs to be roughly the same, if not, market developments can be represented in part of the portfolio and not in another part. Specific factor models have been developed to adjust close prices from markets in Europe or Asia when valuing a portfolio at a US close time.

Time series data that is less frequently available includes macroeconomic data. Numbers, such as unemployment, inflation, and GDP are typically released on a monthly, quarterly, or yearly basis. Financial statements can also be considered as time series data and are typically made available on a quarterly, semiannual, or annual basis.

The concept of time series is spreading beyond prices, order books, and volumes. As we will see in the subsequent text when discussing ratings, banks also have to keep histories of internal credit ratings and operational losses as part of regulatory requirements.

Liquidity is about time and place. If all trades arrive at the same time and place, no brokers or exchanges would be necessary. Liquidity can be defined as the opportunity available to the trading party to implement their trade ideas cheaply and quickly (for an excellent discussion on liquidity, see Harris, 2003). The function of organized trading platforms is to connect buyers and sellers. Different aspects of liquidity include:

- *Immediacy*. How quickly the trade can be executed.
- *Width* or *breadth*. This is the cost of doing a trade, for example, the bid–ask spread plus any commissions. It gives you one of the aspects of the cost of trading, the cost of liquidity.
- *Depth*. This refers to the size of the trade that can be arranged at a given cost; for example, 100 shares will obtain a better price than 100,000 shares, first because the market maker will not have that much inventory, and second because the market maker will suspect that he or she knows less than the person who wants to move 100,000 shares so will sell only that much after adjusting the quote.
- *Resiliency*. It refers to how quickly prices revert to former levels after a large-order flow imbalance initiated by uninformed traders. Uninformed refers to those cases where traders are not acting on new information but merely want to rebalance their position. Different players focus on different aspects of liquidity. Large traders that want to move big blocks focus on depth; impatient traders focus on immediacy and market breadth.

Liquidity is a part of the cost of trading. Frequent trading can erode the gains from the strategy. Together with other costs it goes into transaction cost analysis ("TCA"). Note that decreasing transaction costs is often easier and more reliable than improving portfolio selection decisions and that reductions in trading costs directly affect the bottom line.

Liquidity depends on the type of market and on the instrument and can change very quickly as seen in the 2007–09 financial crisis when the market in mortgage products and commercial paper vanished overnight. Typically, blue chip stocks and foreign exchange for OECD currencies are very liquid. Government bonds can be reasonably liquid; many corporate bonds and municipal bond trades are often in thin supply. Exotic currencies are illiquid. Liquid and active markets have large numbers of investors that come to the market for different reasons and with different investment horizons. You will tend to find more diverse market structures than in less active markets so there will be more order types and market models to choose from when implementing a trading strategy.

For larger trades, the trading costs in terms of direct fees as well as market impact can be high. If institutional investors deal directly among themselves, they could save costs. Systems, such as Liquidnet (see www.liquidnet.com) aim to facilitate this. The system looks at the trades on the blotter and checks if it can match with someone else, if so then anonymous messaging is used to negotiate the price.

Trends in liquidity data include:

- More transparency on trading activity. Transparency used to be confined to exchange traded products. With posttrade reporting obligations under the upcoming MiFID II and already implemented for certain OTC products under EMIR and Dodd–Frank, there is much more information available on trades in, for example, swap markets. This information is collected in Trade Repositories.
- Central clearing of OTC products has led to a standardization in certain OTC markets. This in turn makes it easier to compare and aggregate trading volumes.
- Liquidity is influenced by regulation, such as Basel III. If it becomes costlier to hold securities or if it becomes more important to hold certain high-quality liquid assets (HQLA) under Basel III, supply may shift strongly.
- Regulation increasingly also asks to report on liquidity risk. This goes both for banks and for investment managers. This in turn has led to the development of new liquidity scores (e.g., for bonds that historically trade OTC; for some information on data products see https://www.theice.com/market-data/pricing-and-analytics/analytics/liquidity and http://www.markit.com/Company/Files/DownloadDocument?CMSID=3e73ade9d7c1461091776bd5afefa65d).

2.6.5 Analytics

To make sense of the often very large sets of time series data, different statistics and indicators have been developed. Zero curves, par yield curves, and forward curves are different expressions about the structure of interest rates as a function of length of time. Volatility smiles and surfaces express the structure of the volatility implied in options contracts as a function of strike price and time to expiry. Different data vendors including services, such as Riskmetrics (see http://www.riskmetrics.com/index.jsp) provide market summary information that can directly enter an application. Riskmetrics provides correlations and curves that can directly be used by risk and pricing engines.

Risk factor data is part of the group of derived data: calculated prices based on market data and the terms and conditions of the instruments. Risk factors are summary measures—often the price drivers in a portfolio can be summarized into a small number of risk factors that explain most of the variation. The process of finding those key risk factors is called *Principal Component Analysis*. Other analytical measures include:

- Volume weightings, such as VWAP. This is the volume-weighted average price for a certain time interval. From all the transactions in this time interval the price and volume are multiplied and then summed. The VWAP is the ratio of this sum to the total volume traded within the time period. It is a key benchmark to measure the level of execution of a trade, that is, to see how good the achieved price is.
- Pricing Adjustment. These are different adjustment to correct an observed price for a certain (hidden) risk. Examples include Credit Valuation Adjustment (to lower a price for the credit risk of the counterparty), funding valuation adjustment (FVA), and debit valuation adjustment (DVA, to adjust by your own credit risk). Collectively these adjustments are termed XVA. The trend in valuation is to be more conservative and to apply different *Prudential Valuation* adjustments. On the data sourcing side, the trend is to cast a wider net when it comes to different sources of price information including exchange prices, traded prices both internal and from third parties, tradable quotes, consensus service data, indicative quotes, and counterparty collateral valuations (see, e.g., the standards on Prudent Valuation from the EBA on https://www.eba.europa.eu/regulation-and-policy/market-risk/draft-regulatory-technical-standards-on-prudent-valuation; see also http://ec.europa.eu/transparency/regdoc/rep/3/2015/EN/3-2015-7245-EN-F1-1.PDF, chapter 1, article 3) (Fig. 2.5).

    For risk factor data too, these increasingly need to be supported by real-life transacted prices rather than just models (e.g., in the *Fundamental Review of the Trading Book*; see http://www.bis.org/bcbs/publ/d352.htm, paragraph 183c).
- Different types of corrections are needed to correct the price for nonlinear sensitivity effects, for instance, the convexity adjustment is a correction factor used in the context of interest rate futures and the option adjusted spread is the part of the yield spread of a fixed-income security that is a result of embedded optionality, for example, in convertible or puttable bonds.
- Another category of data is that of factor models that are often used in statistical arbitrage. Factor models are statistical models that represent instrument

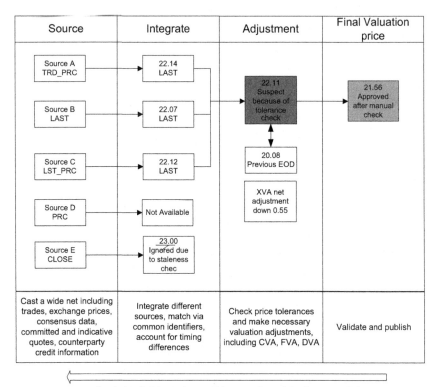

FIGURE 2.5 Valuation process—data management steps. *DVA*, Debit valuation adjustment; *FVA*, funding valuation adjustment.

returns by a weighted set of common factors plus an instrument-specific factor. The weights are called factor loadings and the factors can include macroeconomic variables, interest rate, inflation, credit spreads, stock index level, and volatility.

Many analytical measures express *sensitivities* of a price to a certain variable. We discuss the most common ones for options, equities, and fixed income in the subsequent text.

Options. Option sensitivities are also known as the "option Greeks." These measure the sensitivity in option price to different underlying price drivers. These are the following:

- *Delta.* The sensitivity in price based on changes in the price of the underlying.
- *Gamma.* Sensitivity in price based on changes in delta. In other words, this is the second derivative of the option price with regard to the price of the underlying asset.

- *Rho.* The sensitivity in price based on changes in interest rate with which positions can be financed.
- *Vega.* The sensitivity in price based on changes in the volatility in the price of the underlying asset.
- *Tau.* The sensitivity in price based on changes in time to expiry.

To calculate these measures and to calculate the option price, the option contract reference data, the current market prices of the underlying, and the current risk-free rate for the remaining period of the option are all required.

Equities. The two measures in the equity world are *beta* and *alpha*. Beta is the portfolio's correlation with the benchmark and measures the similarity of behavior with it. Alpha is the surplus return over the index or benchmark. There are a number of ratios typically associated with funds, such as the Sharpe Ratio, Treynor Ratio, and Jensen's alphas that can all be used for fund ranking and comparison purposes. (The alpha/beta paradigm is used mostly in equity portfolios and is also often used in promotional material.)

For equity performance, the effects of the corporate actions that took place need to be taken into account. Time series can be analyzed with or without the effects of stock splits, dividends, and other capital distributions taken into account. Some vendors deliver uncorrected historical time series, whereas others may have applied splits. Since the user requirements may be different, this represents a challenge for common data services within financial institutions.

Fixed Income. Fixed-income instruments typically move in a much more systematic fashion than equities. In fixed income, two measures are normally used to indicate interest rate sensitivities: *modified duration* and *gamma or convexity*. Modified duration is a measure of the true life of the bond; it measures how fast cash comes back to the bond holder and takes the effect of intermediate cash flows into account. It measures the sensitivity of bond prices to changes in the interest rate. Convexity or gamma is the second-order effect and measures the sensitivity of modified duration to interest rate changes. Modified duration is similar to delta for options and convexity is similar to an option's gamma. The data quality of the benchmark yield curve needs to be very good since it determines the value of the bond portfolio.

Commodities. There are many different kinds of commodities and not all of them have an active and liquid futures market. Every type of commodity has specific risk factors as industrial demand and supply dynamics determine the value. One risk measure often used is that of the *basis*. The basis is the difference in price movement between two related commodities, for example, crude oil and jet fuel. Other risk measures are *spreads* that price the conversion effect of commodities at different stages of an industrial supply chain. Examples of this include the *crack spread* to price conversion from gas to electricity or another measure to go from sugar to ethanol. The main data source to measure against will be the most liquidly traded commodity within a group that will serve as the benchmark commodity.

Many data providers also provide risk factors and analytical measures on top of the data. However, the regulatory tendency is for more transparency, meaning firms will still need access to the underlying data and need to be able to reproduce and/or explain the calculations that led to the analytics.

2.6.6 Legal Entity Data and Entity Relationships

Reference data on legal entities—the actors in the high-level financial markets data model that cover customers, venues, issuers, regulators, and market infrastructure—includes different types of information including:

- Basic information, such as the full legal name, the country of incorporation, and the address.
- Classification information on industry sector and client type. Common industry classification schemes include NAICS (see http://www.census.gov/eos/www/naics/) in the United States and NACE (see http://ec.europa.eu/competition/mergers/cases/index/nace_all.html) in the European Union. Clients and counterparties are classified into different types according to fiscal obligation and regulatory requirements, for example, whether they are financial/nonfinancial counterparties (EMIR) or retail, eligible, or professional clients (MiFID II).
- Standard identifiers to identify the party. This can range from social security or ID/passport information for individuals to chamber of commerce information for enterprises. Recently, the Legal Entity Identifier ("LEI"; ISO standard 17442) has been introduced as a standard global identifier for participants in the financial markets. At the moment of writing the coverage is over 460,000 LEIs issued globally (September 2016; see https://www.gleif.org/en/lei-data/gleif-concatenated-file).
- Legal hierarchy information showing the company's legal structure including all subsidiaries, but also information on shareholders, directors, and management. Some entity data products specialize on the cross-referencing between various identifiers and show the linkage between issuers and issued instruments. This can help the client identification within a bank and is needed for credit risk to link issuers to issues.

Use cases of this information start with customer onboarding where a due diligence process has to take place to

- screen individuals, often against a name of politically exposed persons (PEPs) and to determine citizenship to prevent claims by tax authorities;
- screen companies by a compliance department. A firm needs to decide whether it wants to do business with a counterparty and if so, which products and what limits are appropriate.

Second, legal entity information is used when extending credit. This requires a thorough examination of a company's financial statements, business plans, and execution capability and management.

Different data products have been developed to provide information on legal entities. These can be divided into those given in the next subsections.

2.6.6.1   Corporate Structure

Content products on corporate structure show a holding company with all its wholly owned or majority-owned subsidiaries. Good-quality information on who owns who, who owns what, and who is ultimately responsible for any obligation incurred somewhere in the legal hierarchy tree is very important. Whereas many suppliers provide basic rollup information so that you can find the name of the ultimate (domestic or global) parent, other suppliers specialize in the provision of the full structure that allows you to track all subsidiaries from the holding company down. As many larger companies can have well over 1000 subsidiaries that often have similar names, in a structure perhaps 6 or 7 levels deep, this can be a complex picture.

Additional information includes ownership percentage, names of board members, and names of the major shareholders of the holding. The users for such products would be corporate finance departments who need to know the ins and outs of clients they advise, and credit risk departments who need to know how certain credits roll up into one exposure.

Example products include BankersAlmanac from Reed Business that provides specific financial and organizational information on banks and Bureau Van Dijk Electronic Publishing. This company provides different company databases including Amadeus (covering 21 million public and private European companies), Bankscope (providing information on 32,000 banks worldwide), Oriana (focuses on the Asia Pacific region), and ORBIS (covering over 200 million companies) (Source: www.bvdep.com).

2.6.6.2   Industry Classification

On top of the governmental industry classification schemes, such as NAICS and NACE, widely used commercial classifications include the Global Industry Classification Standard (GICS), which is owned by Standard and Poors and MSCI, and the Industry Classification Benchmark (ICB) from FTSE and Dow Jones Indexes. Generally these standards offer a tree-like structure with various levels to designate major industry groups, and subsections down to more precisely label company segments. This information can help group exposure to companies into different industry buckets to monitor concentration risk and can also be used to create industry segment indices and creating a peer group to assess performance.

2.6.6.3   KYC Support

In recent years, the demands on client data management have increased. This is partially due to consumer protection legislation and the increased screening of clients' behavior to detect illegal activities. Measures include Know Your Customer (KYC) that refers to due diligence a bank has to do on a client's background before the client is taken on and Antimoney Laundering (AML) that refers to the monitoring of a client's money transfer behavior. KYC includes checking the client's name against a watch list of known money launderers or criminals. Often these lists are kept by governments and law enforcement agencies, such as the *Specially Designated Nationals* list from the US Office of Foreign Assets Control (see, e.g., http://www.treas.gov/offices/enforcement/ofac/sdn/t11sdn.pdf). AML includes checking the behavior of clients against their anticipated or recorded profile. Specialized software that detects suspicious patterns in payment behavior can flag clients for investigation.

2.6.6.4 Relationships

A specific category of information is that of relationships between the different actors in the financial services ecosystem sketched in the previous data model. This includes relationship types, such as:

- Registers of who's who. Which individuals hold what board positions, and what number of non–executive director positions? What are the family ties? Increasingly regulators keep registers of people active in the financial services industry. It can also include a register of PEPs. The use here is to track behavior and to screen individuals before taking them on as a client, employee, director, or service provider.
- Social network information from LinkedIn or Facebook shows relationships between people in a social graph. Information on online activity of individuals shows preferences and patterns used for targeted B2C advertising.
- Relationships between businesses extend beyond parent–child in a legal hierarchy. Specific information on all companies and their client–supplier relationship can be used to make investments or decide on credit, or to analyze a particular industry.
- Specific financial relationships between actors and objects in the high-level data model include a guarantor relationship if one firm guarantees another firm's debt, an issuer relationship to identify the legal entity in, for example, a bond contract, and different agent relationships including collateral, clearing, and settlement.
- Confidential information on who are the holders of a specific security to gauge market impact, to know who is buying from whom or to connect buyers and sellers.
- These are typically directories containing the names and contact details of fund managers or dealers in various products. These directories can be divided into product types (e.g., overview of currency traders) or company types, such as hedge fund directories. These can be used for sales and marketing purposes and for customer relationship management. Publishers of this kind of information include Euromoney (see http://www.euromoney.com/). It has become increasingly easy to construct your own lists and to group individuals by professional interests using tools, such as LinkedIn.

- Inside a company there can be private social networks as a substitute to email for collaboration projects.

Specific technology developed to represent and track properties of relationships is that of graph databases that we will cover in Chapter 5.

Compared to security master information that benefits from a long history of standardization attempts, complications in the area of counterparty data management include:

- Large variety of sources. A large variety of sources for legal entity data including chambers of commerce, rating agencies, court proceedings, news feeds, direct research, websites, and data aggregators.
- Internally sourced information. A larger proportion of data is sourced internally as some information is proprietary, for example, internal ratings, client profiles (e.g., MiFID customer classification and best execution agreement), and reporting hierarchies.
- Diversity in content licensing. There is commercial content like you have in security master and in streaming pricing information. However, some information is public (financial statements of public companies), other information is sourced peer to peer through business relationships, and other information is confidential (limits, client profiles, holdings, internal credit ratings, and exposures). Contrast this with security master or pricing data that is much more homogeneous. This means specific challenges for access rights and data protection.
- More politics as to ownership of data. Whoever owns the client data is said to own the client relationship. Product master data is more neutral.
- A different data quality process and a lack of standard identifiers makes for a different matching process. The LEI code is changing this and is planned to be expanded beyond just identification to also contain linkages to convey the legal structure. Unlike security master you will not have a lot of redundant sources in use to cross-check vendors or to ensure business continuity: the process will be more about completing the jigsaw by sourcing complementary data sets (financial, research, relationship information). This means the data management process is more about integration than comparison and the matching process is typically fuzzier. A lack of identification standards means matching to link sources will take place by looking at many different attributes (legal form, country of incorporation, sector, string matches) or on internal identifiers.

2.6.7 Portfolio Data

A portfolio can be defined as a collection of positions in different financial products under the responsibility of a specific investment decision-making body. This can be a committee, a board of trustees, or an individual portfolio manager. Every consumer is in this sense a portfolio manager.

On the sell side the term trading book is also used as a unit of accounting. If you own the trading book, you are responsible to keep the risk within limits and trade according to a certain profile. If third-party money is managed, the term portfolio manager is used.

Portfolios can be cast in the form of an investment structure. This can be done directly in a separate account for the client or through the medium of a fund. There are various types of fund that are regulated depending on the level of sophistication assumed in the target client. A mutual fund would fall under UCITS rules in the European Union; a hedge fund would be much more loosely regulated but would not be open to the general investment public. Information on a portfolio includes its:

- Portfolio manager. The person or persons responsible for investment decisions.
- Mandate. What is the investment policy, and what types of products can the portfolio invest in? Can the fund borrow money, is there a minimum cash holding, and so on?
- Operational information. Who values the portfolio? Who administers the assets? This includes relationships with fund administrators and custodians.
- Benchmark. Against which measure is the performance of the portfolio evaluated? This can be a publicly available index or a custom benchmark.
- Legal status. An investment fund can be a hedge fund open only to specific clients or it can be a mutual fund subject to specific regulation.
- Risk scale. Certain fund types, such as UCITS funds in the European Union require a risk scale and Key Investor Information to be disclosed (see http:// ec.europa.eu/internal_market/investment/investor_information/index_ en.htm for contents of the key investor information document).
- Other risk/return scores aimed at consumers provided by Morningstar or Lipper.
- Historical return information and information on any dividends paid by the fund.
- Top investments and/or decomposition of the investment by country, currency, or industry sector.

2.6.7.1 Return Analysis

When analyzing the returns of a trading portfolio or of a fund, the specific return measure and the period over which it is measured have to be clearly defined. Apart from a simple comparing of values at the beginning and the ending of the observed period, there are other return measures including:

- *Risk-adjusted returns* that combine risk management with management accounting. These metrics measure the comparative performance of business lines and portfolios by including a cost charge for the capital used. In this way, consistency is introduced into business unit performance measurement.

Measures, such as risk-adjusted return are good KPIs to see how efficiently the institution's capital is put to work.

- *Risk-adjusted excess returns.* In this case the return is compared to the return of the market multiplied by the beta of the portfolio.
- *Total return.* This takes reinvested dividends into account.

Different considerations and input information as to a return measure include:

- Are the returns gross or net of taxes that need to be paid (e.g., withholding tax)?
- Are the returns gross or net of expenses and fees due to management of the fund?
- What is the return period? Is the return a Year to Date return, is it a rolling 1-year return, or is it the return of the last calendar year?
- What benchmark is used if an excess return is claimed? Has the benchmark been consistent over the life of the fund and over the return period?
- How does the return rank against the risk? What kind of return would be expected for the level of risk taken? In other words: what is the risk unit price per performance unit?
- Is there a guaranteed return, for example, through an options structure? Is it possible to retrieve the price of that option construction or to replicate it via other products?
- Are interim cash distributions from the fund included and what are the assumptions on the reinvestment of those cash distributions?
- Are accruals, such as accrued coupon and dividend payment accrued since the last cash distribution included in the return measure?
- Are there other benefits incurred through holding this investment? If so, can these be quantified and added to the return, for example, income through securities lending?
- How has the portfolio been valued prior to the return calculation? Was accurate market data used? How have its illiquid assets been valued? What valuation methodology is used? Is it mark to market and if so what prices are used? Is it mark to model, and if so has there been independent verification of the models? Accountants and risk managers started to become more aligned after the introduction of fair value accounting in 2006. Fair value accounting stipulates that three approaches can be used to revalue positions (see http://www.fasb.org/summary/stsum157.shtml):
  - market approach through evaluated prices;
  - income approach using discounted cash flows to assess value;
  - cost approach by looking at the replacement costs of an asset.

When using evaluated pricing, it is always possible to revert to a discounted cash flow method if no quotes are available. Normally information, such as quotes, spreads, prepayment speeds and Loss Given Default (LGD) measures is used to come to an evaluated price. After the financial crisis, the focus on fair value accounting and using prices as much as possible anchored in real transactions has increased.

- Have currency effects been taken into account? If so, are accurate exchange rates used?

The fund may have its own benchmark that it seeks to mimic or beat in which case the *tracking error* is the difference with the benchmark. Which benchmark should be picked? Sometimes there is no appropriate benchmark. One person's tracking error is another person's alpha. If the fund has no benchmark, then investors will still have certain expectations on return on the basis of the risk profile of the fund and past behavior.

2.6.7.2 Portfolio Performance Attribution

The *attribution* process of a portfolio checks if the portfolio is appropriately positioned against the mandated risk factors and exposure. Good performance attribution of a portfolio will have both a tactical and a strategic use and will

- use a benchmark that reflects the strategy (and mandate) of the portfolio/ fund (strategic component);
- lead to an understanding of the relative effect each risk driver had to the performance of the portfolio;
- highlight the portfolio manager's skills in asset allocation (tactical component).

Data needs for attribution include the portfolio segment weights and their returns and benchmark segment weights and their returns. Good attribution means high data costs, both for gathering it internally and for collecting the data that is needed externally.

2.6.8 Transactions

Transactions are the specific interactions between actors in the data model. Simply put, a transaction is described by

- A date/time when the transaction takes place.
- Identification of the two parties to the transaction and potentially information on an agent or clearer that will become the middleman. This could also be the place where the transaction takes place if it is on an exchange.
- Identification of the product that is traded and the amount. This would be the number for shares, the nominal amount for a bond, and the notional amount for an OTC derivative.
- In case of derivatives the terms that describe the product, such as strike price for an option or interest index for a swap.

A specific type of transaction is a funding transaction, such as a repo (repurchase agreement) that takes place to temporarily liquidate an instrument. An instrument is sold and at the same time an agreement is struck to buy it back later at a slightly higher price. This comes down to a secured loan.

Specific data standards have been developed to describe transactions to make processing more efficient. Specific transaction standards include:

- FIX and FIXML. This standard is mostly used for listed products (see http://www.fixtradingcommunity.org/).
- FpML. This standard is mostly used for derivatives (see http://www.fpml.org/).

Records of transactions need to be kept for different purposes. For instance, a client may challenge a trade or the price in the trade. Due to regulatory developments, more and more transactional information needs to be reported. Trades that are OTC need to be published to a trade repository in the interests of regulatory market surveillance and to improve price discovery for the general public.

This development started in the United States with reporting rules for swap markets to the CFTC in the Dodd–Frank act and spread to the EU with EMIR rules for OTC derivatives. At the same time, transactional info is required in valuation and risk processes that have to rely less on models and more on real-life transactional info. Different Trade Repositories have sprung up that collect these transactions for the banks and brokers. These include UnaVista, DTCC, Regis-TR, ICE, CME, and Bloomberg. The range of financial products that needs to be reported on is increasing and MiFID II in Europe will lead to additional trade information being available.

2.6.9   Financial Statements

Risk management and finance are at the end of the chain and have to combine different information threads coming from different client segments, products, divisions, and countries. Financial statements are published externally, signed off by external auditors, and therefore undergo specific scrutiny. In addition, since they are meant to inform investors and to represent an accurate view of the business, there are specific rules on how to record the value of assets and liabilities. Many countries used to have their specific principles and rules (*Generally Accepted Accounting Principles*) but there has been global convergence toward International Financial Reporting Standards (IFRS).

Financial statements include the cash flow statement, income statement, and balance sheet. Financial statements are typically produced quarterly and annually and the (market) data that is used at month end, quarter end, and especially year end will undergo additional quality checks.

The trend is toward more rigor in valuation using demonstrable market data. Also, the repercussions of material errors in the statements for chief financial officers and accountants alike have gone up.

The requirements on detail in financial statements are most extensive for publicly traded companies. In the United States they are tracked in the Electronic Data Gathering, Analysis, and Retrieval system (EDGAR) database (see https://www.sec.gov/edgar.shtml).

A standard increasingly used to exchange business information is XBRL. This is used in financial reporting (FINREP) to regulators, such as the Federal Deposit Insurance Corporation and the EBA and the standard used in reporting to tax authorities is XBRL (see, e.g., http://www.xbrl.org/glkeyfeatures/gl_webseminar_lutescohen_051215.pdf for examples).

2.6.10 Risk and Regulatory Reports

Similar to financial statements, risk reports represent aggregate information views on the state of a firm distributed to external parties. Risk reports are much more specialized and provided to regulators instead of to the general public.

Especially since the financial crisis of 2007–09, regulators have become more demanding in the breadth and the depth of the information requested. On the breadth, the trend has been to cover more and more risk categories. Regulatory reporting started with credit risk reporting—the main activity of a bank. Following growth of capital markets and trading activities, market risk was added in the 1990s. This in turn was followed by metrics on operational risk introduced in Basel II. Most recently, risk metrics added are the Net Stable Funding Ratio (NSFR) and Liquidity Coverage Ratio (LCR) to provide metrics on both intermediate- and short-term liquidity risks.

The depth has increased too. On top of day-to-day portfolio risk metrics, regulators have introduced different "what if" scenarios to gauge the effect of different shocks on a bank. This "what if" analysis requirement is spreading to other types of financial firms. Within the European Union, the EBA has been harmonizing reporting frameworks and standard reporting forms for FINREP and common reporting (COREP). This was needed since before each European Union member state had to locally implement the rules and this led to reporting differences.

COREP has become the standard reporting framework issued by the EBA for reporting under the Capital Requirements Directive and covers a number of categories including credit risk, market risk, operational risk, own fund, and capital adequacy ratios. The framework has spread beyond the European Union and has been adopted by almost 30 European countries. The data standard used for the periodic filing is XBRL.

The financial services industry is looking for ways to reduce the cost of regulatory compliance. One approach is to see which part of the reporting process can be shared. In the case of KYC obligations there are verification steps on accounts and new clients. For the case of market or credit risk some of the underlying data could be pooled, for example, enhanced market data services that provide standard sets of risk factors. Possibly the regulators could provide this themselves, similar to the stress tests they supply, to improve comparability in risk reports.

Conceivably regulators could ask firms to open up completely so that they can directly pull the granular position and loan-level data from the firms to run their own reports whenever they want. The technical barriers for this have gone away; the remaining hurdles are behavioral and legal.

An attested submission is probably much better from the regulators' point of view however, and is more understandable from a firm's management point of view as well. What seems more likely is an agency model where the firms pool their more granular data in a shared environment and the regulators pull from that. The advantages of such a continuously updated database with granular data at the loan and instrument position level are a consistent view at all times across the industry where the regulators can always do their own outlier detection and flexibly run their own "what if" scenarios. Conceivably this could reduce the reporting burden on banks but they do need to create a continuous "drip feed" granular reporting stream to a database open to regulatory access.

2.6.11 Tax Information

Every jurisdiction has its own tax code. From a financial instrument perspective the following taxes need to be taken into account:

- Capital gains tax. Taxes that are levied when an instrument is sold for more than it is bought.
- Transaction tax. A tax that applies on a per transaction basis. The famous example of that is the Tobin tax concept that would apply to currency conversions. The stamp duty payable by the buyer of shares is the oldest tax in Great Britain. Recently more countries introduced financial transaction taxes. This is a tax that is difficult to evade.
- Tax on income from securities. This is a withholding tax applied by countries on dividend and interest income. In the European Union the withholding tax is withheld by the country in which a citizen has an account and this tax is passed on to the country in which the citizen is a resident. Increased cross-border sharing of information means that it is harder to avoid these taxes.
- **"Foreign Account Tax Compliance Act"** (FATCA) requires US citizens living outside the United States to disclose their non-US financial accounts and also requires non-US foreign financial institutions (FFIs) to report the identities of people with US person status in their accounts to the US treasury. Each financial instrument has a FATCA status (Yes, No, or Grandfathered) and reports identities of such persons and assets to the US Department of the Treasury.

Due to increased crackdowns on tax evasion, banks need to know more about where there clients are taxed and in the case of FATCA face a heavy burden to determine US person status of their account holders. Due to international pressure, countries, such as Switzerland have relaxed their banking secrecy laws. Tax sharing between sovereign countries is increasing, although some countries still collect very little information on shareholdership of companies reasoning that they cannot share what they do not collect.

The operative term in tax information is Ultimate Beneficial Owner—the person who is behind a corporation and benefits from a certain structure. Tax havens typically do not disclose who the shareholders of a company are.

2.6.12 Other Documentation

Contract information refers to the legal framework trading parties put in place and includes agreement on settlement terms. The settlement risk can be reduced by agreement between two parties to have all their dealings in a certain product type (e.g., credit derivatives) governed by the ISDA master agreement. The ISDA master agreement came out in 1987 following the growth of OTC derivatives markets and was amended in 1992 and then again in 2002 following the market crises of the late 1990s.. The master agreement covers different product categories and for different geographical areas and determines the basic framework within which two parties trade.

Under a master agreement, confirmations of specific trades fall under the provisions of the master agreement. This will, for example, imply a *netting agreement*; transfers between two parties will involve only the balance of all the sums due for all the trades under the master agreement. *Transactional netting* means parties can net out amounts payable on the same day and in the same currency. The scope of *close-out netting* is broader. In this case, when a master agreement is cancelled, all the transactions under it are valued to determine settlement amount. These settlement amounts are converted into a termination currency and added up; the result is one net payment from one party to the other.

Contract data also includes the administration of which products can be sold to which customers. This includes both exposure limits to cap the risk of the institution and also the set of "eligible" products for retail clients, high-net-worth individuals and professional counterparties. Increasingly, financial institutions have to take proper care that they offer only appropriate products to and commensurate with a client's risk appetite and investment objectives. In the case of professional counterparties, contract information can also include details on "best execution" agreements that stipulate, for instance, the eligible execution venues for a transaction plus other execution criteria.

Other than contracts that govern the trading relationships between two parties, each security has its own legal documentation in the form of a prospectus. The prospectus can be very long depending on the complexity of the security. In Europe, legislation has been introduced to improve investor protection by stipulating that all prospectuses provide sufficiently clear and comprehensive information. The other objective of this regulation is to make it easier for companies to raise capital throughout the European Union on the basis of approval from a single member state authority.

Other than the prospectus, **research reports** are issued after issuance of an instrument and give the opinion of a broker on companies or specific products. Analysts cover a certain industry sector, track a number of companies, and provide advice on whether to buy or sell a stock. Research reports are created by brokers to generate orders and unsurprisingly most of the advice is to buy something.

Earnings estimates and fundamentals are important data categories. Public companies have to periodically file their financial statements, such as income statements, cash flow statements, and balance sheets. Information on past performance is therefore in the public domain and is disclosed in different ways. The United States has the EDGAR database and SEDAR is the Canadian equivalent (see http://www.sec.gov/edgar.shtml and http://www.sedar.com/). In some other countries the chamber of commerce keeps registers of companies and financial statements.

Many analysts forecast earnings and dividends. Forecast information on different financial measures, such as revenue, cash flow, dividend, earnings per share, and accounting terms, such as EBITDA is collected. Companies, such as FactSet (see http://www.factset.com/, e.g., their Lionshares database) offer access to multiple underlying databases for financial information and analytics through one front end. This also includes information on ownership of equities.

Other research information is provided by media companies that provide, for example, industry sector reports that look at the state and future of, for example, the oil industry, the steel industry, and so on. This type of research is done both within buy and sell side firms and through specialist research companies.

2.6.13   Communication Logs

Communication logs include data captured from emails, phone recordings, chats, and social media activity. Recorded one-to-one interactions preceding a transaction are an example of the increased retention and usage of *unstructured data*. The point is to prevent market abuse, protect investors, and provide a mechanism to retrieve the environment in which trades came about. Dodd–Frank requires all oral (phone, voice mail) and written (email, chat, text messages, fax) communication that led to trade execution to be retained. For a single trade this could be a number of chats over Bloomberg plus a recording of a confirmation phone call. Also, earlier legislation including Sarbanes–Oxley included email retention policies. Banks need to track the communication records of their individual traders much more closely following the rate fixing scandal in LIBOR.

The point of regulation (including MiFID II) is to keep a record of the **context** surrounding a transaction. Both the structured information, such as the counterparty and trade identifiers and the unstructured records, such as voice communications, electronic messages, and social media are kept. This means associating an enormous amount of information with specific transactions— plus the data not only needs to be stored but also needs to be organized and retrievable. Compared to Dodd–Frank, MiFID II impacts a broader group of market participants and employees.

There are implications for traditional document management systems and new document databases to allow for efficient archiving, retrieval, and mining of this information. We will discuss this in Chapter 5.

Social media information is not only tracked to prevent market abuse but also an important source of information on retail customers. Consumers increasingly volunteer personal information that can be used for marketing products (Facebook, Twitter) or for finding new staff (LinkedIn). For financial services firms, social media can be an extension of their branding. APIs on LinkedIn or on Twitter provide mechanisms to tap into these information streams. Twitter occupies a space in between (informal) news and social media and is a useful source for sentiment analysis (see Firehose on https://dev.twitter.com/streaming/firehose).

2.6.14 News

News is provided by agencies that can be commercial wire services, corporations, and cooperatives, such as newspapers that pool their content. Financial news on macroeconomic data, key hires, financial results, patents, client wins, and the outcome of lawsuits is actionable information and a major driver of prices. Services, such as Factiva (see http://www.dowjones.com/products/product-factiva/) provide add-on services to make the classification, processing, and distribution of news content easier.

The International Press Telecommunications Council (IPTC) (see http://www.iptc.org/pages/index.php and see https://iptc.org/standards/newsml-g2/ for information on the standard) has created a standard for the formatting of news called NewsML. NewsML aims to facilitate the formatting and smoothen the supply chain within news that is composed of agencies, editorial systems, news aggregators, and users. The standard includes metadata, such as status information (a press release could be embargoed) and copyright information. Other than this, the standard allows for the identification of various multimedia types of content, such as images and video. When used in archiving news, it should make querying and retrieving information easier.

Standards, such as NewsML help to feed scanning software that analyzes news. This can be simple, for example, to scan for keywords, such as company tickers to put a news item in the right bucket or to make personalized web pages but also, more interestingly, to actually parse a piece of news, to be able to understand that it is an earnings statement, to filter out the key numbers of a company press release and to then compare it to analysts forecast, or to detect the positive or negative in a news statement. The automatic interpretation of news and also the ability to trigger investment decisions is a next step. Protocols, such as rss help in filtering public sources on the internet to create a custom news feed to a desktop.

Apart from the formatting of news to facilitate interpretation, vendors of news could also *flag* news stories to indicate interesting content, possibly based on client instructions. Apart from this, they can also offer *news archives*, which clients could analyze if they research, for instance, price behavior around the breaking of news stories. Special news services exist that collect, for example, information on Mergers and Acquisitions, for example, Zephyr from Bureau van Dijk (see www.bvdep.com). These kinds of services provide up-to-date information on M&A activity, venture capital deals, IPOs, joint ventures, and private equity deals.

News is increasingly moving from flat text to marked-up documents to facilitate machine processing. Scanning services try to read the prevailing mood to gauge sentiment and to build investment signals. Simultaneously web crawling services, such as owlin.com track large sets of sites and news services and combine this with machine translation to harness a large array of local language media. Local media often pick up information earlier. On top of machine-readable news and crawling services, dashboards track information pickup or build trading signals or warnings on anything that may impact an operation (supply chain, supplier prices). Services, such as Twitter are an intermediate form between news and social media.

Personalized news dashboards that crawl the web for you and distill information and key statistics relevant for each user are the norm.

2.6.15   Credit Information

Credit ratings are assessments of creditworthiness of a retail or corporate client, of an issuer, of a financial instrument, such as a bond or structured product, or of a country. They are either provided by research agencies through research on behavior, financial analysis, or statistical analysis or created within a financial institution (internal rating). In this section, we distinguish between the following types of ratings:

- credit ratings
- mutual fund ratings
- commercial ratings
- retail credit scores

Each of these four categories is discussed in the subsequent text.

2.6.15.1   Credit Ratings

Large rating agencies, such as Moody's, Standard and Poors, and Fitch dominate this market. These three agencies all have Nationally Recognized Statistical Rating Organization (NRSRO; see also http://www.sec.gov/answers/nrsro.htm) status in the United States, which means that their ratings can be used under SEC regulations. Smaller agencies include start-ups or specialists in certain product sets or geographies through deep knowledge of the local market (agencies specific to Japan, such as Mikuni and JCR, Indian agencies, Canadian Bond ratings from Dominion). Credit ratings are normally based on accounting and macroeconomic models and are positioned as "through the cycle," that is, they are not a mark-to-market snapshot of creditworthiness but presumably a relatively stable and consistent assessment of credit that should last through a business cycle.

Credit ratings are assigned to issuers, issues, and countries (sovereign ratings). Typically the issuer will pay to receive a rating as a rating will help investors to assess the product and will make it easier to sell securities. The rating scales go from AAA (or Aaa depending on the agency) down to D (default). Several other scales are in use to rate specific company types or products, for example, the Financial Strength Indicator for banks. Rating agencies also keep watch lists that indicate whether an issuer or an issue is likely to be downgraded or upgraded in the near future. The agencies have made their ratings products more granular, often providing both long-term and short-term ratings, for the local currency and for the foreign currency. *Composite ratings* present an average of various ratings, similar to a consensus estimate for financial results forecasts.

Credit ratings are used in product pricing and in credit risk management to assess the credit of an exposure but also as a selection criterion on where to invest. Certain funds are allowed to invest only in products above a minimal rating, typically BBB on the rating scale that is the boundary between *investment-grade* products and *high-yield* products.

Ratings agencies provide various add-on content products to help banks in their credit risk process. These products include the provision of a history of each rated issue or issuer and *transition matrices*. Transition matrices show the probability of an issue or issuer moving from one rating class to another within a certain time frame. The approach of assessing credit risk in such a way is also called an *actuarial approach* because it rests on compiled historical data. Alternatives would be to create an analytical model to simulate migration between credit bands.

Over time, rating agencies have moved from research-driven to more market-driven ratings. These developments include the Market Implied Ratings from Moody's that incorporate signals from the credit and equity markets and financial statement analysis, as well as information used in the "traditional" ratings (Source: www.moodys.com). Other measures of credit that are directly priced in the market are through instruments, such as corporate bonds, Credit Default Swaps, and using the equity price data [now part of Moody's Analytics, assessing the credit risk for public companies through valuing the equity as an option on the firm using the Merton model was pioneered in the KMV model of *expected default frequencies* (EDF)]. These developments have led to a continuum in credit assessment between research through macroeconomic and financial statement analysis and dynamic market information.

2.6.15.2 Mutual Fund Ratings

Mutual fund ratings aim to guide investors through the enormous amount of mutual funds available both by providing a classification of funds into styles [e.g., Morningstar through their Stylebox (see http://news.morningstar.com/pdfs/FactSheet_StyleBox_Final.pdf)] and by rating mutual funds within a certain peer group on their past performance. This can, for instance, be done by handing out a number of stars as a result of a statistical analysis on past performance.

2.6.15.3   Commercial Ratings

Commercial ratings are also company ratings but are aimed not so much on solvency but rather as an indication of supplier risk. They provide an indication of the time companies take to pay their bills. Therefore, they can be used by a treasury department or to predict working capital needs. They are not used just by financial services institutions but more typically by companies with a very large amount of (corporate) clients. The major provider of commercial ratings is Dun and Bradstreet (see http://www.dnb.com/us/). Dun and Bradstreet also provides other information on companies and has the largest company database in the world. Their company identification number, the Data Universal Numbering System (DUNS), is one of the de facto standards in company identification and is required for any company that does business with the US federal government. The rating from Dun and Bradstreet is called the Paydex score.

2.6.15.4   Retail Credit Scores

Apart from ratings of corporate clients, banks also require an indication of credit worthiness for their retail customers. There are companies that compile credit reports on private individuals and companies that provide credit scoring models that can be used within banks to rate private individuals. Major suppliers include Experian and Fair Isaac (see http://www.fairisaac.com/fic/en and http://www.experiangroup.com/). Many countries also keep national registers of credit exposure through loans and credit cards that can be consulted by banks and other lenders prior to granting new credits. Credit risk on mortgage payments can be proxied by arrears information. Privacy laws have to be carefully watched in all these cases. The most commonly used credit score is the FICO score developed by Fair Isaac.

2.6.15.5   Internal Ratings

As the pricing of credit is the core business of a bank, the following could be asked: why is it left to an agency? First of all, an agency rating is an independent assessment and therefore valuable as a second opinion. An analogy is in revaluing an investment portfolio where a company will normally not use the trader's own price but an independent price instead. Second, agencies fulfill other roles as well. Their credit stamp on an instrument often has regulatory implications. For example, it can make the difference between an instrument and a structure being eligible for investment by many pension funds and mutual funds.

Over time, internal ratings that banks used to assess credit and price loans have been more formally recognized by regulators as sound risk measures. Under Basel solvency rules, banks can calculate capital requirements on their banking book doing their own assessment of credit factors. Basel II rules allow for different approaches to measure credit exposure. The first approach is the *Standardized Approach* that relies on external ratings only. The second approach is the *Internal Ratings–Based* (IRB) approach where the bank can use its own credit ratings to measure credit risk. This approach comes in two flavors called the *foundation* and the *advanced* approaches.

Ratings help banks keep historical records on the Probability of Default (PD) that they need to calculate the LGD (the part of an asset written off in case of a borrower defaulting). LGD depends on the specific loan since it also depends on collateral and seniority of the loan. The resulting EAD is the Exposure at Default. Different credit risk approaches in the Basel rules require different amounts of historical information. For the IRB Advanced Approach, 7 years of historical loss data is required to estimate LGD, and 7 years of historical exposure data is needed to estimate EAD. Five years' worth of historical data is needed to estimate PD in the Foundation Approach of IRB. In the United States following the expansion of the stress testing program through the DFAST program (see https://www.federalreserve.gov/bankinforeg/dfa-stress-tests.htm) many banks that were not subject historically to Basel rules had to keep historical credit information.

The experience of the 2007–09 financial crisis showed there is a fine line between well-researched credit opinions and public endorsement of a financial product. Therefore, rating agencies have come under increasing scrutiny. ESMA has become the single supervisor of credit rating agencies within the European Union and in the United States it is the SEC. New requirements include disclosure of information on past ratings to help investors make more informed judgments.

2.6.16 Miscellaneous

As more raw material is produced and becomes available, there will always be room for new content products. People pay for content; noise is free. The more noise is produced, the more valuable is the ability to discern the signal.

The market for content is developing rapidly. New vendors occur in established content types but as financial product development and changes in new execution venues take place, we also see new types of content products. The increased availability of information in the public domain including personal information via social media, spatial information via Google maps, and the increased options to tap into large data sets will push new analytics and content product. This could include product development in:

- *Legal feeds.* Content products that, for example, cover all court decisions in a certain jurisdiction. This can be on a retail basis to give information that affects a personal credit scoring (bankruptcy, divorce) or company rulings on, for example, new concessions, licenses, or anything else that can materially affect a share price or a company credit assessment. Currently, this information could be constructed to some extent by filtering news feeds but it could perhaps be obtained more directly. A data model to categorize court rulings would be needed to categorize information quickly so as not to be overwhelmed and to rout information easily and speedily to users.

- *Property feeds.* A content product that provides all the property deals in the land registry. This can include a change of ownership, partitioning of a lot, new allotments, and information on transactions. Again, if this information can be filtered on cities and certain planning permissions, such as commercial real estate versus residential, it would increase the value of the offering. This kind of information could be used in the nascent market in property derivatives.
- *Weather data.* More and more information on temperatures, rainfall, and wind speeds is made available, both by national weather bureaus and by commercial meteorology companies. This can not only be used for weather derivatives, such as the CDD and HDD futures but generally for any commodity trading strategy. Temperatures determine, for example, natural gas consumption and agricultural output.
- *Mortality tables.* This information is available but not easily disclosed. Longevity information is needed for life insurers and for derivatives, such as longevity swaps. Recently indices have been defined in this space, for example, by the Life and Longevity Markets Association (see http://www.llma.org/).

In the cases mentioned previously the information is already there but the content products are not yet fully up to speed to allow for electronic processing and easy integration with other data categories. Investor needs and regulatory changes will continue to drive new content product development.

2.7 CONCLUSIONS

The earlier sections have covered content that is either publicly available or obtained from vendors. In the case of proprietary data, such as transactions and holdings there are other considerations in the supply chain. Whereas in commercially sourced data, you have to abide under the content licensing agreement, in proprietary data there may be confidentiality agreements with clients. "Chinese walls" separation between different departments imposed by regulators, and privacy laws canrestrict unauthorized access to the information.

In this chapter we have taken a closer look at the taxonomy of financial information. We looked at different perspectives to segment the landscape and gave an overview of the data creation processes and the role of the various actors in that chain. We also provided an overview of the different types of content sets. We have looked at dynamics in the content market and at the different types of content financial institutions typically source externally. New players and products come up and generally—because of the lengthy and costly information supply chain—there will be more demand for products that can be more easily deployed. With the rising need to collect and publish info, new content products will continue to be developed. Key trends include:

- From a content perspective, the products need to become more accessible in terms of using industry standard identifiers or providing cross-references between the most commonly used keys of instruments and issuers.
- From a technical perspective, the products need to become more accessible in the sense that they should become easier to integrate either directly with applications (potential for vertical integration for software and content vendors) or with an organization's distribution infrastructure, such as middleware.
- From a user perspective, they need to become more tailored to the actual data requirements. This means a tuning capability in terms of both what information fields are received and also offering a more granular selection on instrument types and markets. Products have become much more "pay as you go" rather than one-size-fits-all solutions. We see this trend already with more and more "portfolio"-type content products launched where a client submits his or her instrument universe to the vendor that then keeps the information on those selected instruments up to date.

REFERENCE

Harris, L., 2003. Trading and Exchanges—Market Microstructure for Practitioners. Oxford University Press, Oxford.

Chapter 3

Information as the Fuel for Financial Services' Business Processes

Chapter Outline

A Primer in Financial Data Management. http://dx.doi.org/10.1016/B978-0-12-809776-2.00003-X
Copyright © 2017 Elsevier Ltd. All rights reserved.

3.1 STEPS IN THE INFORMATION SOURCING PROCESS

Financial services is the business of information processing and risk assessment. All financial services can be summarized into a small number of primary processes and still a small number of secondary, support processes around those (Fig. 3.1).

Primary processes include:

- Trading: buying and selling financial products as a principal
- Investing: buying and selling financial products as an agent, often in the form of a specific entity, such as a fund and against preagreed boundary conditions on risk and eligible assets
- Lending: the supply of credit in the form of consumer lending, mortgages, SME lending up to larger bank loans
- (Re)Insuring: underwriting risks through offering insurance policies
- Intermediary agent services: raising money through selling and issuing securities (bonds, equity) for corporates
- Payment services
- Distribution of financial products: through agent or branch network distributing investment, savings, retirement, and insurance products

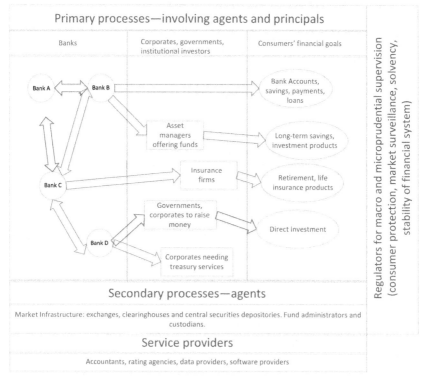

FIGURE 3.1   Different actors in financial services.

Secondary processes include:

- Exchange/execution venues: organized processes to bring buyers and sellers of financial products together
- Clearing and settlement: services to act on the details of a trade and ensuring that both parties get what was agreed
- Custody and securities services: managing securities accounts of firms, tracking lifecycle events against instruments held (this can include ancillary services, such as payment agent, withholding agent for tax and agency roles in securities services, and collateral management)
- Fund administration: valuation, domiciliation, and processing of investment funds
- Internal risk management and internal audit as well as financial services regulation

Regulation spans the primary and secondary processes—but in most cases does not extend to the service providers.

In this chapter we will discuss the information collection processes, seen from the perspectives of a supply chain as well as from the perspectives of the life cycle of the main components of the data model in the previous chapter. Increasing amounts of ready-to-process data are publicly available. Importantly, whereas industry participants used to have to go out and find it—carefully collect it, prepare it, curate it, and place it in predesigned databases for analyses—these days it is often volunteered. However, without proper attention, the quality of data required for daily business processes will degrade over time, often leading to a situation shown in Fig. 3.2.

Changes in data processing include:

- There is a lot more informal information available—also caught unstructured, because it does not live in the formal confines of a predefined table structure in a relational database.

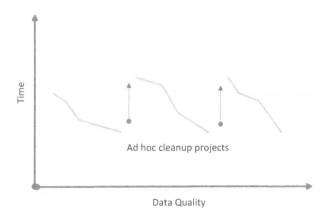

FIGURE 3.2 **Data quality seesaw: gradual degradation and periodic cleanup.**

- The boundaries between unofficial and official are moved back as more unstructured data can, despite its name, be processed.
- An implication from the previous point is that data no longer needs to be in a formally designed relational database to yield insights (Fig. 3.3).

The business of financial services has always been about finding, defending, and keeping an edge in information collection and processing, whether this is about using sources of information ignored or inaccessible to others, seeing the value in information ignored by others (leading indicators, correlations, macrotrends), or having an operation that can distill decisions out of this information (processes, people, computing power).

Large sets of data previously hard to find are now readily accessible to all consumers and customers of financial services. They too have access to a lot more computing power, the internet that lifts the veil, and thousands of apps and applications to collect, process, and compare information and financial products. As a consequence, the bar for financial services to add value through an information edge is raised. Businesses used to be *data constrained*; currently many businesses are at risk of drowning in data.

The value add in selecting the data for decision making is an exercise much like the selection processes used by publishers. Availability of data and democratization of information processing through end-user tools, such as Excel empowered users but also led to a vast new wave of information sources and information uncertainty. It led to each individual user being an end point for further information processing and to an unmanageable collection of spreadsheets. There is a risk of exacerbating that situation now with the big data tools that promise to be implemented quickly, with a very fast learning curve. The difference is now that end users not only get control over their own calculations but also get enormous power in business intelligence that can lead to confusion when there is no clear data infrastructure. One of the primary objectives of information management is to put all users and applications on a common footing, to provide shared data services. The collection and screening of information is more important than ever. End-user applications always constitute a trade-off between flexibility and control. Shared data services lower the potential for data discrepancy and will make end users and end-user applications more productive.

3.1.1 The Information Supply Chain

Most financial institutions have multiple commercial content providers. Especially for larger financial institutions, there is not one main content provider that can satisfy all the information needs. Second, for business continuity purposes and to not depend on one vendor both operationally and commercially, institutions often opt to have multiple sources for the same information. Third, by comparing and commingling content from multiple information providers, discrepancies and errors can be caught and corrected. An example of this in the case of mingling various price sources to produce an approved revaluation

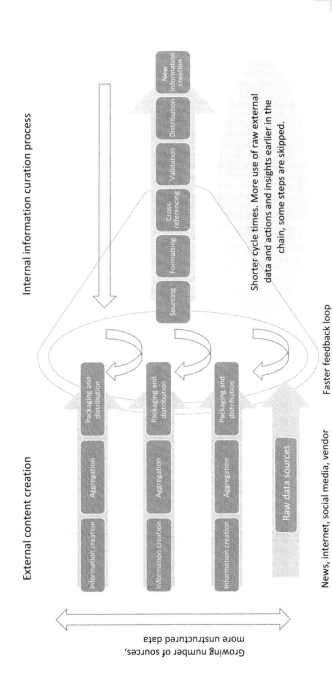

FIGURE 3.3 **Information supply chain.**

price is provided in the subsequent text. The ability to trace the process back upstream is typically a critical requirement. Data creation and sourcing within the firm is needed to build a complete picture.

Data from multiple sources is combined for the purposes of business continuity, to plug the gaps in coverage, and for quality improvement. The internal information supply chain usually includes the following steps distributed over multiple applications and departments:

- *Sourcing* of the data from the providers. This includes manual entry of data through copying from faxes, emails, face-to-face client conversations, web pages, and newspapers. The automated scrubbing, scraping, or shredding of web or broker pages is also part of the information supply chain. This step includes retrieving the information from the web, pulling it from an ftp server, calling a vendor's API, making a phone call, scanning a prospectus and parsing an email.
- *Reformatting* the data into a common format. Often an organization has standardized on a certain naming convention, has defined its own XML structure, or relies on a standard from a vendor of one from the public domain. Reformatting the data can involve "Extract, Transform, and Load" (*ETL*) tools: a set of products that specialize in manipulating and reformatting data, XSLT tools that can transform one XML standard into another. A whole cottage industry around Enterprise Application Integration has sprung up. Often a lot of proprietary and local information needs to be part of the configuration of these tools so although they help to solve the problem, they are just part of the answer.
- *Standardization.* The common format to which data is mapped can be an internal proprietary standard or an open (ISO) standard. There are standards for the *identification* of instruments (ISIN code, ISO 6166 REF) and legal entities [Legal Entity Identifier (LEI) code, ISO standard 17442], for the *classification* of instruments (CFI code, ISO 10392) and industries, such as NACE in codes in the European Union (see http://ec.europa.eu/competition/ mergers/cases/index/nace_all.html) and NAICS in the United States (see http://www.census.gov/eos/www/naics/). There are standards to describe financial transactions (ISO20022, FIX, and FpML) and attempts for complete financial data models (for a good overview of these standards, see McKenna et al., 2014). Standards tend to be less widely used as they get more comprehensive. Identification and classification standards are widely used, transactional standards find pockets of use (equity transactions for FIX, fixed-income derivatives for FpML), and comprehensive financial models are rarely used completely. At some point in the scale of comprehensiveness, standards move from being an enabler to being a constraint.
- *Matching, relating, or linking* different sources for the same instrument, corporate action, and legal entity. This can vary from exact matching on common identifiers, such as ISIN code and place of trade for listed instruments to more "fuzzy" matching for legal entities where you have to check on the

same name. Comparing company names often includes text comparison and text manipulation including expanding abbreviations and stripping away legal form markers, such as "Ltd," "SA," or "Inc." In the case of corporate actions, matching is more straightforward and typically takes place on a combination of fields that define the event type, effective date, and the identifier of the underlying security.

There are also commercial sources that act as aggregators of identifiers and provide this cross-reference information. The creation of these products is complicated since there are many vendor proprietary identification schemes in use; hence agreements need to be struck up with the owner of each code included in the cross-reference scheme. Many new information sources are unstructured and not organized in a structured way around a commonly known identifier or classification standard. This poses challenges to data integration. Intelligence comes from the ability to draw analogies and integrating and deriving insight from data that previously existed in isolated silos or different data sources is valuable. Increasingly accurate pictures of client creditworthiness can be constructed based on profiling social media behavior. This offers alternatives to classic retail credit scoring.

- *Data integration and consolidation.* Combining the content from various providers becomes possible once they have been matched. Blanks can be filled in as vendors often provide complementary information, and at the same time discrepancies can be detected. For example, vendors can disagree about certain information or one can be later than the other in supplying a piece of information.
- *Validation.* This can include formatting checks (correct structure and check digit of the ISIN code), consistency checks (does the maturity date lie beyond the issue date), and price validation checks (deviation against index). Validation often takes place with a four-eye methodology that is required for certain data types in some jurisdictions. In this case, one pair of eyes does the actual correction or validation action and a second pair of eyes signs off on the work. For validation, workflow tooling is frequently used that groups quandaries about related issues together for easier routing and processing and distributes data issues over staff members, also for load balancing among staff. Issues can be organized by priority, by portfolio, by (internal) client, by owner, and so on. The approved data is often called *master data* or *golden copy data* and still needs to be delivered to end users and applications.

An almost philosophical point is whether data constitutes truth or opinion and, moreover, whether these two categories can always be distinguished. We can separate facts (everything that can be tied to official records, such as published financial statements, court rulings, terms and conditions from prospectuses, exchange close prices) from opinions (valuations, credit assessments, research, chats, texts, emails). Factual data changes infrequently and is often found in the reference data discussed in Chapter 2. Opinions are everywhere and change in the light of new factual data – opinional data is often time series data. Sometimes the line is blurred, for instance when there are interpretation issues around published statements. Data integration can sometimes shift opinion to fact, clarify the boundaries, or build a more complete picture of the different opinions out there.

When talking about sources of data, this can mean something fundamentally different depending on the category. The word "source" for factual data often refers to the distributor or carrier of the information. Here is normally only *one* ultimate source: the company that created the financial product. In the case of opinional data, such as credit or quotes, the source is normally the quoting party or trading venue. In this case, multiple sources are *expected*. Also note that when there are multiple "sources" for reference data, the expectation is that they are the same; if they are not, at least one source is wrong. In the case of opinions, the sources can be the same but will generally give different values reflecting different quoting times, different trading strategies, and product inventories of market makers. This means that consolidation over multiple sources of reference data is a matter of comparison and for time series is much more a statistical exercise.

3.1.2 Technical Challenges

The content industry has struggled to address the sourcing needs of their clients. On the one hand, selling directly into desk-level applications may be in a content vendor's commercial interest short term. On the other hand, vendors have made their content products more flexible, technically in the sense of more retrieval and processing options (portfolio-style products allowing selection on specific instruments and attributes) and commercially (per use pricing).

The quality in terms of well-formedness and ease of integration of commercial data products varies enormously. The ultimate test for a data vendor's product quality is often to see how easily it can be integrated and automatically processed. The general trend is that data consumed shifts upstream. The traditional data supply chain and curation process has been shortened. Data is captured earlier and plays a role in decision making sooner.

3.1.3 Data Deployment

After the data has been approved, it has to be brought to a place where it can be acted upon. This can be loading the information into the application where it plays a role or where it can be seen, queried, inspected, and changed by an end user. This we call data deployment.

Traditionally, a key difference between client and proprietary information on the one hand and vendor and public data on the other hand is how it is stored. Data that is sourced from public and/or commercial sources is often stored in a separate data store decoupled from reporting or business applications that may need it. It is often abstracted out of applications and is common to a variety of business functions.

Proprietary and client data on the other hand is usually stored at the application level itself. This includes risk management, portfolio management, or banking systems. Coupling data with applications implies a far lower degree of standardization in terms of data model, taxonomy and semantics.

Cloud-based storage of master data and transactional data has changed the dynamics of data flows. New security and privacy concerns on data outside the firm's own data center have led to increased checks and controls on data flowing between internally or externally hosted systems. We will explore different data management tools and techniques in Chapter 5.

Approved data can be put onto a distribution layer that routs it to the applications that need it. This can be done on a publish subscribe model, where updates to a certain data universe are pushed through to the using system, or it can be done on a request reply–type basis. Any financial services institution should carefully measure and control where information is distributed to and how heavily it is used, to satisfy business requirements and to conclude the appropriate content licensing agreements, but also to be able to attribute the cost of the information and its processing in an equitable way to the business owners. Cost attributed is cost controlled and treating information as overhead will lead to overstated requirements.

In business applications, incoming data leads to actions and decisions (cash transfers, financial instrument orders, dividend processing, client portfolio revaluation, withholding tax calculations and regulatory reporting). New information is created that can in turn become the starting point of the next information supply chain as it is published externally to business relationships or regulators or internally to fuel other processes. The effectiveness with which an institution can produce new content determines its success. Also note that any data sourced from outside a firm is the end product of a data curation process on the side of the data supplier (Fig. 3.4).

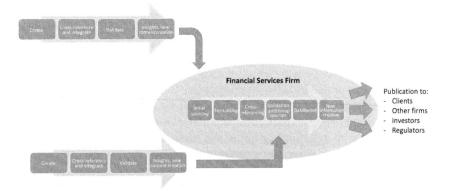

FIGURE 3.4   Intrafirm data collection and curation process—sources can come in at different points.

Generally, this supply chain holds for all commercially, internally, relationship-based, and publicly sourced data. Sometimes, some of the steps are skipped or combined in one. Client or peer-to-peer retrieved data is that which is sourced on a piece-by-piece basis through a business relationship. Proprietary data is created as a result of internal business processes and includes internal ratings, client profiles, pricing models, and the institution's own trading portfolios. In this case the overall picture still holds but there is less automation and standardization. More is captured and more can be analyzed earlier.

The time delays that can occur as a result of a lengthy curation process can be highly impactful. The delays incurred in each of the steps are becoming smaller as the process itself becomes more integrated. The steps we see internally (cross-reference, identification, error correcting) also occur to some extent at each of the steps in an external part of the supply chain, that is, the curation processes at the side of the data suppliers. The trade-off between speed and quality will be different depending on the business function serviced. To take two extremes: for real-time data speed is of the essence. The trader will know the market level and not be too bothered or distracted by the odd rogue quote. For regulatory reporting, speed is less important and quality of information is critical as penalties and reputational damage can be immense. The trend is to **condense** the curation process, helped by technology and data integration. Simultaneously, more and more sources enter the supply chain. New models of consumption based on pay-as-you-go commercials and increased flexibility to cater to consumption-driven demand have come up (this includes companies, such as Xignite and Quandl; see www.xignite.com and www.quandl.com).

3.2 DATA MANAGEMENT FROM THE INSTRUMENT LIFECYCLE PERSPECTIVE

Financial products are created continuously. These range from different products that will be listed on public markets to products created for specific audiences all the way to an audience of one. We discuss the different ways of issuing a new product and subsequently discuss the servicing aspects of those products until the point when they reach their end of life.

New instruments are being created to fulfill funding, investment, and risk transfer needs. The life cycle of a financial product covers many stages with different service providers active at every stage.

The conception and creation of instrument takes place at exchanges, at product development groups, at origination departments that are the middleman between corporate funding needs and investors, and bilaterally between professional counterparties and financial institutions. Instruments are traded on exchanges, on crossing networks, and over the counter. Trading is facilitated by central markets, such as traditional exchanges and new execution venues, including organized trading facilities and multilateral trading facilities under MiFID II, An MTF is a multilateral system operated by an investment firm or market operator that brings together multiple third party buying and selling interests. OTFs are non-equity broker crossing systems.

3.2.1 Issuing

Securities, such as equity and bonds are issued to bridge the funding needs of companies or governments and the investment needs of investors. Typically banks are involved to structure the security and to price it. Initial Public Offerings are when a company first publicly raises equity. These securities come with their own unique prospectus and legal documentation. In some cases there is a master legal structure, such as a Commercial Paper or Medium-Term Note program that governs a series of debt issues.

Listed derivatives, such as futures and options are created by the exchange on which they are listed, such as CME, Eurex, or ICE. They are created to generate turnover for the exchange and to appeal to investment or hedging needs of the users of these products. Listed derivatives are issued against a calendar, typically on a quarterly cycle. The specific terms and conditions as well as how the set of strike prices of an option are determined are covered in the rule book of the exchange.

Products, such as swaps, forwards, swaptions, and exotic options are created through bilateral agreement between the two parties in the form of a contract. Parties often trade under a master legal agreement (see http://www.isda.org/publications/isdamasteragrmnt.aspx). Mortgages and other loans are created continuously and find their way to a broader investment public through securitization. Eventually listed instruments retire through expiration (futures and options), maturity (bonds), or delisting (equity). OTC products retire as per the contractual conditions between the two parties.

3.2.2 Asset Servicing

Following the instrument's issue, there are different maintenance activities tied to the holding of a security. We label these activities as *asset servicing*. These services include *custody* (the safekeeping of securities and the processing of corporate actions), *securities lending* (the temporary transfer of ownership against a fee), and *proxy voting*. These services have extended beyond administrative services to make sure the owners of the securities receive any benefits (coupons and dividends) toward more value-add yield enhancement services (lending the bonds and stocks against a fee), risk management services, and securities financing services (collateral management). Securities services either can cost you money (through custody fees) or can make you additional money (through securities lending operations).

Apart from regular income activity on securities, there can be unexpected *corporate actions* too. These can include bankruptcy or M&A but also some types of bonds can be redeemed early at the issuer's discretion ("callable bonds"). Securities are monitored by Central Securities Depositories (CSDs)

TABLE 3.1 Different Activities and Corresponding Data Needs

Business function	Activity	Data needed	Data quality focus
Custody	Notary function	Unique ID of securities, beneficial owners	Beneficial ownership records
Asset servicing	Process income out of products, voting instructions, securities lending	Corporate actions, beneficial ownership, securities lending eligibility	Corporate actions, timely processes, prices for collateral management
Fund administration	Valuation and tax services	Valuations and tax matrix	Correct prices, correct holdings, tax implications

and Custodians for the banks who in turn pass on the information to the beneficial owners of those products. Similarly, futures and options also need monitoring. Exchange-traded options can be exercised against the seller of a call or a put and corporate actions can also impact futures and options (Table 3.1).

We discuss the various activities in the area of asset servicing, such as custody, securities lending, collateral management, and fund administration below.

3.2.3 Custody

As the name suggests, custody refers to the safeguarding of securities in a vault, or nowadays, when most securities are *dematerialized*, the safekeeping in secure systems. In addition to this, global custody can comprise many other services including:

- income collection and other corporate actions management (this can also include proxy voting services);
- cash management funding and other banking facilities for the account holders;
- tax management through expert knowledge of different fiscal regimes;
- reporting through periodic statements on accounts and holdings.

In addition to this, custodians also offer services that include:

- investment accounting via the tracking of cash inflows and outflows;
- securities lending and collateral services (through securities lending operations, custodians can enhance the yield on the assets the account holder has put under the administration of the custodian; the securities lending fee will be split between the security owner and the custodian);
- trustee services;
- portfolio valuation and performance reporting.

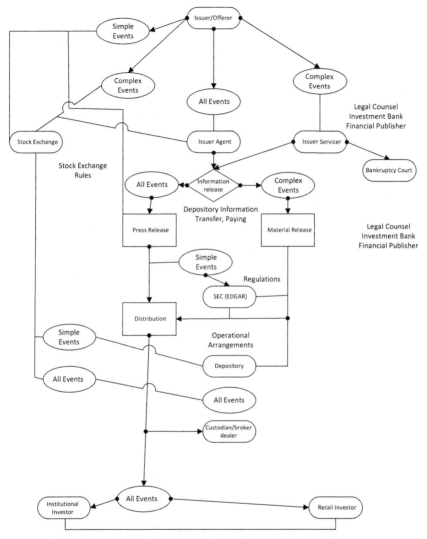

FIGURE 3.5   Corporate actions communication flow. Drawn from https://xbrl.us/wp-content/uploads/2010/12/20100630CorpActionsBusinessCase.pdf.

Corporate action processing still leads to lots of operational losses with risks including (Fig. 3.5):

- direct risk of processing failures, such as choice dividends and rights;
- direct costs of late payments via interest accrued on late dividend payments, potential cash flow problems;
- risk of suboptimal trading decisions by the front office if information is not quickly disseminated;

- indirect cost of ineffective corporate governance through a faulty proxy voting process;
- costs to reputation and costs of reconciliation through systems and staff.

The trigger for corporate action operations is often a SWIFT message from the central securities depository (CSD), or a custody department. The effectiveness could be improved by making this more into a more *proactive approach*, for instance by creating an alert if you do not receive information when you expect it. Sourcing this data closer to the source would make a lot of sense. The burden could be put to the issuer, requiring the company to publish its announcements in a standard format to data aggregators or through the SWIFT network.

Custody and corporate action processing solutions typically cover the following corporate actions processing functionality:

- integration of corporate actions from various sources, typically custodians and commercial data vendors (this includes data collection, scrubbing, and the creation of a golden record);
- entitlement calculation and reconciliation of calculated amounts due versus those actually received based on account holdings information and events, including tax tables and cost basis calculations;
- formatting and submitting response instructions (MT565), confirmation/entitlement payments data (MT566), and confirmation of responses (MT567);
- workflow queues to organize and prioritize messages, facilitation of replying to elective events, and other events, such as claims and class actions;
- capabilities for custodians and account holders to give instructions on how to treat events via a portal.

Corporate actions processing solutions are essentially data integration services that combine internal and external information with a workflow to organize the responses. Benefits of these solutions are a reduction in operational risk and increased efficiencies in processing. Operational risk control should lead to less fails and claims and potentially also a lower operational risk charge for those institutions governed by the Basel Accords. The targeted workflow identifying and prioritizing items in need of attention should make staff more productive, hence reducing costs per trade. As custody is typically charged in basis points per portfolio value, this is a way to increase the operating margin for custodial business.

3.2.4   Case Study: Information Issues in Corporate Actions Handling

Corporate actions have long defied automation, because, just as in the case of OTC derivatives, there is a lot of variation. Common data problems include:

- Corporate actions that are discretionary actions, such as tender offers. In this case there is a need for follow-up and the custodian needs to secure a response from the owner of the securities. The information supply chain is an interactive process that needs to come up with a response before a certain deadline.

- Embedded instruments where a corporate action takes place on an instrument that is underlying something else that is traded. This could be a warrant, an index, or something else. A separate internal announcement on the derivative that is affected by the event on the underlying would be needed. The structure of the derivative product needs to be clearly defined and linked to the corporate events on the underlying.
- A need to *identify and track security family trees* and what happened over time to a security as a result of name changes, capital events, mergers, spin-offs, renaming, and new identifiers. This can mean integrating the corporate actions data with a legal entity data product.
- If a company trades IBM shares in Germany and in the United States, there will be delivery on two places. Instruments that are fiscally treated differently have different ISIN or CUSIP codes and often you will want to aggregate per ISIN code.
- A choice has to be made on the granularity of the data and the processing model, for example, whether processing is based on an instrument or on a listing level. Representing events on an instrument level means a requirement to have more depth to support multiple ex-dates. (Some exchanges may have a bank holiday on the ex-day.) Eighty-five percent of all corporate actions are plain vanilla and can be represented in a flatter model not based on the listing but based on the instrument.
- The existence of multiple "fraction dispositions." Let's take a 2 for 3 split and a current position of 5; this can be mapped to a new position, which will be 7.5; then the question is how to treat the 0.5 (these fractional shares are also sometimes called *script*).

    The issue is that most vendors look only at the fraction postsplit and not any fraction presplit. Similarly, some information of layouts can be region or country specific. Expressing ratios of splits can differ: a 2 for 1 could be expressed as either a swap 1 for 2 or have 1 get 1, 1:1 or 2:1, respectively.
- Events can change or be filled in over time, for example, a provisional dividend is followed by an exact dividend amount, or an additional option can sometimes be added. This requires tracking the history of an event.

Given the variety of corporate actions, attempts to standardize can only go so far. Where there is a limited number of large brokers, imperfections in data standards can be addressed by additional process agreements and on how to use the standards (see the guidelines on corporate actions market practices from the Securities Market Practice Group on www.smpg.info).

3.2.5 Securities Lending

Securities lending is a temporary transfer of securities on a collateralized basis (this discussion draws from Faulkner, 2008). The word "lending" in the term is misleading as it is not a "loan" at all, but a title transfer. The duration of this "loan" can be on-demand or term. Many lenders want to preserve flexibility and loan equities on call. The economic benefits go to the borrower. However, the borrower typically "manufactures" these back to the lender.

For securities lending, the securities need to be negotiable. In the United States you can hold securities in your own name or in street name; in the latter case the broker or the DTC holds them in trust and the beneficiary owner can recommend the broker how to vote. Transferring securities into *street name* means making them negotiable.

Motivation for securities lending includes the following cases:

- *Shortfalls.* Securities lending is often triggered by a failed settlement in an original trade that has led to a shortfall. Settlement needs to be faster as there is normally a short-term need to have these securities to cover a shortfall.
- *Yield enhancement.* This refers to the case when a portfolio has been constructed with borrowed securities that yield more to the borrower than the costs that need to be paid back to the lender. The *carry* is the interest return on the securities held minus the financing costs.
- *Tax arbitrage.* In case two sides of the transactions fall under two different tax regimes (one side may receive tax credits), it can be more advantageous for one side to receive a dividend or an interest payment (due to a difference in withholding tax regimes).
- *Index tracking.* Another interesting motivation for securities lending lies in index tracker funds. There are ETFs that have to track an index very closely. They cannot take stock dividends since this would mean a deviation from tracking the index accurately. In this case they can lend the equity, have the borrower receive the stock dividend (assuming the stock dividend is economically the more attractive option compared to cash dividend), and get it back with a cash return higher than they would otherwise have received.

The distinction between securities lending transactions and repurchase agreements ("repos") is fuzzy. Generally, securities lending deals come about as a result of a need for a specific security (ISIN, CUSIP) and repurchase agreements are to fulfill funding needs. So it is also a question of what drives the transaction; what may be a repurchase agreement to one side of the deal may be a securities lending transaction to the other party.

Typical information to be included in a securities loan transaction includes:

- transaction and settlement date
- term/duration
- security identification
- security price and quantity
- loan value and lending fee
- collateral and margin (top-up amount)

Note that day count conventions differ in the United States and the United Kingdom, so the basis for interest calculation is important to determine.

Securities lending used to be more of a back-office to back-office activity but has evolved into a profit center. It is frequently outsourced to third-party securities lending agents. As the volume of securities lending transactions has increased, there are more intermediaries who pool together supply and demand. Sometimes there are agents in between who specialize in finding the right securities, just as any other broker activities but now to arrange a securities lending transaction. These providers offer various services including counterparty selections, matching long and short portfolios, and providing negotiation platforms.

3.2.6   Collateral Management

In OTC transactions, often there will be master agreements on settlement and netting in place. These include provisions on collateral pledging to secure an exposure. Eligible collateral between two parties needs to be clearly defined as well as the procedures to replace or top-up collateral and the size of the *haircut* or discount applied. There is a tendency to move beyond the relatively crude way of valuation of a collateral portfolio through a haircut to more precise NAV calculations. This implies that good-quality pricing data is needed here as well.

The size of the haircut will depend on the volatility of the collateral, the proportion of the total security issue held in portfolio and the liquidity. There will be agreement on the absolute value of assets to be accepted as collateral, the initial margin and the *concentration limits*. This refers to the maximum percentage that a specific asset, all assets issued by the same issuer or any asset class can represent in the total collateral pool. Multiple bonds and equities can be pledged but if they all refer to the same ultimate credit, there may be an undesirable level of concentration. These limits need to be monitored against up-to-date information as the market can quickly turn or dry up. Eligibility criteria on what constitutes good collateral have gone up and monitoring is taking place more scrupulously.

Market and credit data needs to be up to date and accurate for good collateral management. Following the financial crisis, there is less appetite for unsecured lending and collateral management has flourished. Collateral management includes the management of liquidity risk, mispricing risk, and legal risk. If there is a possible delay in selling the collateral securities, then the risk is higher. When collateral quality deteriorates because of adverse market conditions, it needs to be substituted or topped up. Triparty collateral agents, such as Euroclear, Clearstream, or JPMorgan Chase offer services on top of their custody services since they already know what assets depositories have.

3.2.7   Fund Administration

There are many activities that need to be performed to support the investment process. The fund administration function has evolved from different specific services into a full-service back-office function that includes daily P&L and NAV calculations. Fund administration is frequently outsourced and typical activities can include the following:

- *NAV calculations.* NAV calculations include dealing with all capital inflows and outflows within a fund, fund income and fund expenses, and the maintenance of the fund's financial records. We also call these activities *fund accounting.* [For a fund it is better to keep expenses low (so limit the hard dollars) and to get all services included in the commissions. Commissions are not expenses but depress investment returns. Under MiFID II, research has to be unbundled and is separately chargeable.]
- *Risk management services* in operational risk as well as fund benchmarking and attribution. This means that fund administration providers have a need for good benchmark data.
- *Reporting.* Investment products can be offshore or onshore, cater to retail or institutional markets. This can mean a potentially complex tax reporting. In addition, fund administration providers may also offer services around a stock exchange listing of a fund.
- *Bookkeeping* for all portfolio transactions: trade and settlement date, date of receipt of broker confirmation, and allocation of transaction volume to different brokers based on settlement quality.
- *Fee calculation*: management fees, performance fees, and other expenses.
- *Compliance*: investment policy compliance, mandate checks, and controls on eligible securities and markets.
- *Administrative functions,* such as the maintenance of the shareholders' register for the fund and transfer agency function.

3.3 DATA MANAGEMENT FROM THE TRADE LIFECYCLE PERSPECTIVE

In this section we outline the transaction life cycle, the processes around a trade that need the content described in Chapter 2. A transaction is an exchange of cash and/or financial products between two parties and leads to a change in the assets and liabilities of the two parties. There are different types of transactions:

- The immediate real-time delivery versus payment (DVP) settlement, for instance, in the US Fedwire for money transfer and FX transactions via continuous linked settlement (CLS) Bank.
- Securities settlement typically has a lag of a few days. Many markets have 3 business days, some 1 or 2 between the transaction data and the actual exchange.
- Many derivative transactions have multiple "lifecycle" events that require settlement, such as loans with a start and an end settlement or swaps with periodic payments that can stretch out over 30 years.

Transactions can take place by firms acting as a *principal* or by firms acting as an *agent*. In the first case the firm uses their own money and will carry the risk. In the second case the transaction is made by an investment manager who has been given a mandate by investors to manage their money.

We discuss what category of data is important where, at which stage it enters the transaction life cycle, and present the most important criteria for success for data in different processes in terms of speed, accuracy, completeness, and control. The processes around the transaction are depicted with various levels of granularity so that individual process steps can easily be seen in the perspective of the complete activities of a large financial institution. This way it becomes clear how activities, such as research, trading, settlement, risk, and reporting are closely interrelated in information flow terms.

The evolution of financial product types from the classic dichotomy between debt and equity to a continuum has put enormous stress on the core processes of a financial institution and has led to a corresponding variety in technologies. Both content and software application product offerings seek to address specific niches and to provide institutions with a specific edge.

The distinction in transaction process between the high-volume/low(er)-margin business of standardized products and tailor-made low-volume, higher-margin products affects information management. The former typically follows a highly automated system with automatic means for price discovery, settlement, valuation, reporting, and so on. The second type of process relies much more on manual work by specialists and on smaller, niche applications that are often developed in-house.

In this section, we will start out with a discussion about the *information manufacturing* that goes on within institutions. We then look at the main steps in the transaction life cycle: how do transactions come about, how are they executed, and what happens afterwards. We will do this both in sequential order—covering pretrade, trade, posttrade, as well as reporting on risk and performance—and through a short description of every application type found running these processes.

We will take the lifecycle approach and address a transaction from end (two parties strike a deal and agree on terms) to end when final cash flows are exchanged and both parties no longer have any liabilities arising from the original transaction.

The main steps we will describe in this section are the following:

- *Pretrade.* This includes the generation of trade ideas, the price discovery process, short-term and long-term research, and product control. On the applications side this includes (pre) trade support systems, such as analytics toolkits, quoting and market making systems, and research applications.
- *Trade.* How are trades executed? Who is the counterparty? What are the various trading strategies, execution models, and execution venues? We will discuss the information needs for different trading strategies, latency, order types, and liquidity. On the applications side, this includes Algorithmic

trading, Execution, deal capture, and Order management systems. Other information needed includes where to clear or settle, which legal framework to use for the transaction, and whether the exposures one party has on the other needs to be collateralized.

- *Clearing and settlement.* What is necessary to make sure that cash and financial product change hands safely and timely? We will discuss trade matching, various central counterparty services, and barriers to integrated clearing and settlement that exist in the industry. From an applications perspective, we will cover posttrade systems including settlement systems/Electronic Trade Confirmation/trade matching systems, payment systems, and ownership transfer.
- *Risk, reporting, and performance management.* This includes activities, such as portfolio valuation and portfolio analysis, risk reporting, compliance, and financial reporting. From an application perspective, it includes portfolio management systems, Net Asset Value calculations, fund accounting, attribution, and performance measurement applications. On the risk side it includes systems for liquidity and cash management, collateral management, market, credit, operational risk, internal rating systems, accounting, and a general ledger system.

We can split up the trade cycle in three parts to arrive at the following breakdown:

	Function	Data requirements
Pretrade	• Quantitative modeling	• Research; time series; reference data
	• Broker/trader discussions	• Counterparties; costs
	• Management approval	• Limits; P&L histories
	• Account management	• Accounts; counterparties
Trade	• Execution	• Execution
	– Order management	– Algo parameters; brokers, EMS model
	– Order routing	– Reference; counterparties
	• Margin and securities lending	• Inventory; availability; margins
	• Account management	• Accounts; counterparties
Posttrade	• Clearing and settlement	• Transactions; standing settlement instructions; accounts
	• Corporate action processing	
	• Risk management	• Positions; corporate actions accounts
	• Compliance management	• Counterparties
	• Account management	• Positions; prices; risk parameters
	• Fund management	• Positions; regulations
		• Accounts; counterparties; positions; prices

This table shows that different types of content will be required in the pretrade (idea generation, price discovery process), trade (execution), and posttrade (settlement, asset servicing, revaluation, risk reporting) steps.

Functions of an organization have traditionally been split into Front, Middle, and Back-Office lines. Because functionality from one area to another has

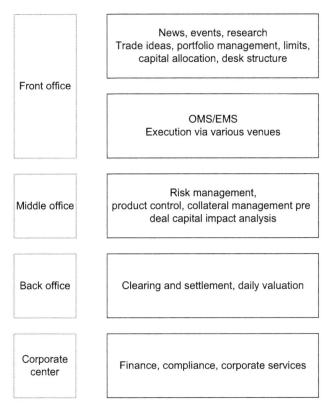

FIGURE 3.6   **Front, middle, and back-office breakdown in a financial institution.**

become more integrated, this division is not all that useful anymore. Outsourcing and offshoring have on the one hand led to a decoupling of functions, often across different entities. On the same page, many financial institutions offer services in one or more traditional back-office activities; they have in fact made back-office services, such as revaluation, processing, matching, and custody into separate lines of business (and thus into new front offices). A schematic breakdown of activities is provided in Fig. 3.6.

Data is extremely important in the whole life cycle (Fig. 3.7), indeed, so much so that financial institutions can be said to be in the financial information data management business. This is also reflected by the IT budgets found

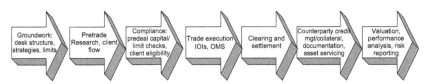

FIGURE 3.7   **Transaction life cycle.**

in financial institutions (total IT spending for the worldwide financial services market for 2015 is forecast at \$461.4 billion and expected to grow to \$534.7 billion by 2018). All application areas and business processes mentioned have their own unique data needs depending on the complexity of the product processed. Information needs vary in terms of the following:

- *Frequency* of delivery: varying from streaming data for pretrade to rubber-stamped end-of-day prices for NAV calculation and market risk control.
- *Granularity* of the data sets. From a reference data perspective, a trading system would need a few attributes to identify and price the security, whereas the custody and portfolio system would need to know all the ins and outs of cash flows and potential corporate events. From a pricing data perspective, an order execution management system would need to have all the level 2 information (quotes from all the dealers or the order book from the exchanges) to intelligently rout the order; a collateral management system would need just the last prices to revalue the pledged securities.
- *Size* of the universes delivered. Applications in different business lines seek information on different products but within a business line different users will need different sets of securities. The settlement and portfolio system will need information on actually traded securities or products in which the institution has an open position. A trading or research system may track a much larger group of eligible financial products. A custody system will need detailed information on all instruments held in custody for every single customer.
- *Preferred sources*. Different end users could have different preferences as to the vendor they want. The branding element of content should not be underestimated.

Note that this breakdown in steps in the transaction life cycle is a simplification. There are often many extra steps and complications that greatly depend on the specific product and market. We need to only look at the enormous variety of information and software products available in the market to appreciate the complexity and the number of niches. Continuing specialization and competition fosters new niche areas in which firms can set themselves apart. This in turn creates new niches for products and services.

The traditional silo nature of the organization charts of financial institutions is to some extent mirrored in product offerings. Products offered by content and software vendors compete for budget held by product line–oriented businesses. Hence, a silo budget structure leads to silo-like product offerings and silo-like licensing agreements. The task of putting in place lateral links is typically an afterthought.

3.3.1 Pretrade

Pretrade is about coming to trading ideas, investment, or trading strategies and getting the groundwork in place to be able to do trades. On the information gathering and research side, this can be to decide upon a trading strategy that best reflects the company's resources in terms of human capital, clients, risk appetite, capital availability, horizon, geographical presence, and other infrastructure. On the execution specifics it includes decisions on market timing and specific instrument selection. Putting in place all the boundary conditions for trading includes funding of trading activities, a mandate in terms of product, capital, counterparties, and risk limits.

In pretrade, we distinguish between the following activities:

- *Idea generation and research.* Research groups within banks and investment managers can cover macroeconomic and country-level research (sovereign ratings, business climate, default risk) that can be used for country, asset class or sector allocation of funds. Apart from strategic research, the research function can also include more focused research groups that do cheap/dear analysis of certain parts of the bond market and that produce very specific and immediate trade ideas for the own front office or to pitch to clients to generate order flow for the bank.
- *Product control.* When an institution creates a new structure or an exotic derivative that they want to sell to institutional clients or think up a new fund for the retail side, it has to go through an approval process. This is the role of product control, which is typically part of Risk Management. The job of product control is to check whether the risk factors in the product offered are properly understood and properly priced. It can also include determining eligible clients for this product as part of a marketing strategy.
- To *simulate* returns, large historical data sets of prices can be required to run sample portfolios and to backtest trading strategies.
- *Limits* are set by a credit committee or by trading operations management to guide the trading activities and to control exposure to, for example, a counterparty and a category of counterparty (by country, by industry sector), and by risk driver (such as exposure to the US stock market, the Japanese stock market, or a specific currency). A benefit of using a central counterparty (using central OTC clearing services) is that this works around limit constraints and the credit risk of one counterparty will no longer be a constraint for business.
- *Legal trading groundwork.* Part of the pretrade process can also be the establishment of a legal framework with counterparties. This includes completing documentation including master agreements, compliance documents, such as Qualified Institutional Buyer (rule 144a) document, clearing agreements and repurchase agreements. Some organizations have a document control function that checks if all the paperwork is in place. This can also include checking *corporate resolutions* that tell you which individuals are authorized to act on behalf of a company.
- *Operational groundwork,* such as getting the administrative details right, including the sourcing of settlement instructions, basic data, such as names and contact details, and creating a record in the counterparty database. This

can all be done posttrade but it is more efficient to do it ahead of commercial activities.

- To start a trading operation, *working capital* to do the trades is required. In this case, there is often a link with an institution's treasury department. Treasury includes activities in cash management and short-term hedging and can also be deriving data to mark futures and swaps contracts and to monitor liquidity risk management and funding costs.
- For *price discovery*, the pretrade process differs also by market. In OTC market, prices can be found on bulletin boards, by bilateral quote requests or IOIs through counterparties, or by using a broker who aggregates quotes and who arranges trades. In some markets, the price discovery process is outsourced to central execution venues, such as exchanges. In a quote-driven exchange market there are roles, such as market maker/dealer who have to continuously quote two-way prices and who are the counterparty to every trade. When there are enough buyers and sellers, an auction market model can function and there is less need for someone to "make the market," since it already exists. In case of a large deal flow that comes to the institution, trades can also be *crossed*; they are arranged between the institution's clients.

Application types found in the pretrade stage include analytics providers and analysis tools but also general reporting tools typically used for longer-term research. Excel and streaming data analysis are used for shorter- to ultrashort-term research. When low latency matters, traditional databases where information first needs to be stored to disk before it can be exposed are not suitable. Database technologies determine the query/information retrieval speed and thus latency. An arms race has taken place in low-latency that has resulted in the collocation of data centers with the exchange to get the shortest possible path between order and execution.

In the trading process there is a distinction between proprietary trading (employing the institution's own capital and seeking revenue from market movements) and acting on behalf of clients (seeking revenue from client commissions). The size of proprietary trading has gone down following the financial crisis; the cost for a bank to have a trading book has generally gone up due to increased regulatory capital requirements.

In both cases information on the trading counterparty is needed to facilitate the processing including SSI data and basic data. In the latter case, information on these customers needs to be in place, including:

- Regulatory information. This includes the initial screening of a client as well as subsequent client behavior screening (AML). This can be both online in real time before the transaction or payment goes through or by retrospective analysis.
- Policy information. This includes a set of products the client is allowed to trade, a set of venues that are eligible execution venues as per the client's instructions (more venues could mean better prices but will also mean

higher trading costs), discretion in order execution, the mandate when the institution is investing on the client's behalf, trading limits, and a list of trading authorized persons from the client, possibly detailed by product and amount.

Several regulations aim to make the price discovery process more transparent but especially to make sure that counterparties obtain the best price available in the market at that time. Both Dodd–Frank in the United States and EMIR and MiFID2 in the European Union have extended market transparency.

3.3.2 Trade

For front-office decision support and the generation of trade ideas, a lot of time is wasted through inefficient browsing of information. Easy access to data is one of the largest untapped productivity improvement opportunities left to financial institutions. Highly paid staff spends too much time on mundane data collection work, especially for more time-critical trading strategies where the opportunity costs of *not having access* to data are often literally prohibitive. New data sources and BI techniques can create new sources of information that can influence trading or investment decisions.

In the case of arbitrage, a hedge portfolio needs to be constructed. For example, in index arbitrage you would have the index on one hand, and constituents (or a major subset of the constituents) on the other hand and you need to keep prices up to date on both. Easy access to related instruments can improve productivity and insight, for example, the ability to link from an equity directly to all the options. For fixed-income securities, look up other Euro government bonds around a certain point on the curve. Comparing the credit spread of bonds against other measures in the credit market, such as CDS can generate trading ideas.

3.3.3 Trading Styles and Data Needs

The implementation of a trading or investment strategy leads to orders. Orders are generally triggered because of conditions that occur that make implementing the trading strategy meaningful. This can be because (in the case of arbitrage) the price of certain instruments reaches a certain level or because (in the case of event-driven strategies) news is announced.

An investment strategy is typically constrained by the mandate of a fund. Boundary conditions can include the risk level, borrowing levels, and markets and products to invest in. Investment strategies typically start with asset allocation and then go down to selections at the instrument level. The question of "where do I want to be invested" explains most of the return.

Data needs differ depending on the trading style. Arbitrage or relative value trading involves instruments with similar risk factors that are correlated. Often the strategy will be about isolating a certain risk factor, for example, in equity index arbitrage, buying the index (future), selling (the large weighted) constituents, for credit trading, buying a corporate bond, and selling the government bond future to isolate the credit spread.

In case of a directional trading strategy, research, news, and any other information about the direction in the risk factor to which you are seeking exposure is critical to fully form your view. Unlike relative value trading, big losses can be incurred in case the factor moves against you and you cannot timely close out the position. From a regulatory capital perspective, this is also much more expensive. You will seek exposure to those risk factors where you believe to have a competitive advantage.

In case of statistical arbitrage, very large histories of tick data are needed, for example, to create market impact models and for simulation. If by contrast you are a classic stock picker, you will study market fundamentals plus the financial statements of the company and will not require real-time data.

Statistical arbitrage is about the screening of quote streams. Event-based trading is about the scanning and interpreting of newsflow. Although these strategies have historically been less time-critical, tentative steps toward the automation of newsflow interpretation have been made. Parsing of news can be about scanning the news for keywords or—when the news story has been tagged—financial statements could automatically be constructed out of the news story and be checked against the consensus forecast.

3.3.4 Clearing and Settlement

A trade *clears* when both sides are in agreement and have a common understanding on all the terms of trade. For example, for a repo transaction both sides need to report the same notional amount, repo rate, length or term of the repo, the security repoed, and the price of the bond. This agreement can come about bilaterally as a result of comparing trade confirms, or clearing can be done by an intermediary utility. A trade *settles* when the product purchased is exchanged against cash. A trade clears if both sides report the same terms of the trade, else it is called an outtrade or a don't know (DK).

Clearing and settlement are two different activities. Settlement is normally on a DVP basis in which the security changes hands against cash. Occasionally, there is the option for settlement through "delivery by value" (DBV) where instead of cash immediately, securities are delivered overnight and cash the next day. Settlement activities include the input of settlement instructions, the verification of trade details, notice of execution (NOE), and transfer of ownership. Reference data is required to support the accurate identification of securities and counter parties Settlement instructions have been discussed in Chapter 3. They tell the other party that the money and the securities need to go. Settlement instructions need to be periodically reviewed to check their accuracy. This can be done through a *twilighting* process whereby counterparties are asked to reconfirm their instructions. They would have the choice between confirming the validity of the existing data on their profile and offering new information. In case no reply comes, the data can stay the same with a note that the client has been asked or individual follow-up can be taken.

More and more products that used to clear bilaterally are now undergoing central clearing (see, e.g., the SwapClear service from London Clearing House on www.swapclear.com). Central clearing has led to increasing standardization of derivative products and is becoming a standard component of good risk management. Clearers step in to be the buyer to every seller and vice versa and, provided they are sufficiently capitalized, form a multilateral safety net for settlement risk. This reduces the risk profile of each individual counterparty, and, as a result, lowers the risk of a settlement failure. Central clearers also support brokers and other agents by connecting buyers and sellers that do not have liquidity to settle trades themselves. Clearinghouses play a key role in the stability, efficiency, and resilience of markets.

Some products have also gone on-exchange. On the settlement side, an important development is the standardization of capital markets within the European Union under the T2S project (https://www.ecb.europa.eu/paym/t2s/html/index.en.html). T2S introduces a common settlement platform and is rolled out over the different CSDs in four successive phases or "waves." The final wave is scheduled to end in September 2017.

Securities and OTC products have potentially very different settlement processes. In securities there is typically a National Numbering Agency that assigns unique identifiers, such as the CUSIP, ISIN, or SEDOL codes, which makes for (relatively) easy processing and identification. This is also the case for structured products that have CUSIPs and settle like a corporate bond. In derivatives, you have to have the legal framework/master agreements established and settle the exposure periodically, for instance via payments through SWIFT.

The CLS bank has been set up by a number of large FX trading banks for the clearing of foreign exchange trades (see http://www.cls-group.com/Pages/default.aspx; CLS currently settles in 18 currencies).

Large clearinghouses include the DTCC, ICE, and CME in the United States and Eurex and LCH.Clearnet in Europe. Some clearinghouses clear only equity or only options instruments. Other clearinghouses, such as LCH.Clearnet offer a wide variety of services in equity, exchange-trade derivatives, commodities, and OTC derivatives. Sometimes, clearinghouses are separate entities, and sometimes they are integrated with an exchange (Clearstream and Deutsche Boerse are examples of the latter category; Euronext and LCH.Clearnet are examples of the former category). In the United States there are services provided by the Federal Reserve including the Fedwire Funds Service for larger amounts, Fedwire securities services, and the National Settlement Service.

Some securities transactions can be internally cleared within a bank; in this case they are offset against another transaction and the bank does not have to go to the exchange or the clearinghouse to execute the trade. Similar to pretrade transparency regulation, there is a regulatory push to posttrade transparency.

The aim here is regulatory oversight of market activity rather than helping investors with price discovery. Some of the services that existed already to rout trades to different places [e.g., Traiana (ICAP), MarkitServ, and Thomson Reuters Trade Notification (TRTN, http://thomsonreuters.com/en/products-services/financial/trading-platforms/trade-notification.html)] can also be used to send trades to a reporting service.

SWIFT plays a critical role in facilitating settlement:

- Through the provision of a safe network to exchange information. This also includes the provision or member-administered closed user groups ("MA-CUG") where subsets of SWIFT members can exchange information reliably and securely.
- Through the provision and acting as the registration authority for different standards including the ISO 15022/20022 standards as well as identification standards, such as the Bank Identification Code (BIC). On the payment side, SWIFT has the BIC directory that can be used with country and branch extensions to identify a bank's branch office. It is also the start of the International Bank Account Number (IBAN) that uniquely identifies an individual bank account.
- Through services, such as Accord for matching and exception handling in foreign exchange, money market, and OTC derivatives.

3.4   DATA MANAGEMENT FROM THE CUSTOMER INTERACTION PERSPECTIVE

An effective information collection and curation process is not only important for trading and investment processes but also critical to support customers. Information trends in terms of the possibility of instant access, a larger number of information sources, and more options to set up data curation processes also impact customer expectations.

Fnancial services firms face a tough challenge in the postcrisis world. Not only has their image been dented, but also they risk being disintermediated in important areas by new firms in the fintech space. Large social media and online retail firms have reset the expectations of accessibility and communication.

Customer interaction with financial services firms has changed dramatically—most of all in the retail space. The role of a bank or investment manager in the life of a client has changed. Gone are the days of the branches. After being redirected to call centers and after branches were marginalized, customers are now used to apps that easily and instantly give them access to information any time (Fig. 3.8). Clients expect to be able to transact anytime and anywhere and financial services firms need to be open for business 24/7. Up-to-date information on all of a client's interactions with that firm needs to be instantly available in a secure way (well described in King, 2012). Retail customer loyalty is no longer rooted to personal interaction with bank staff and switching is easier. Loyalty is

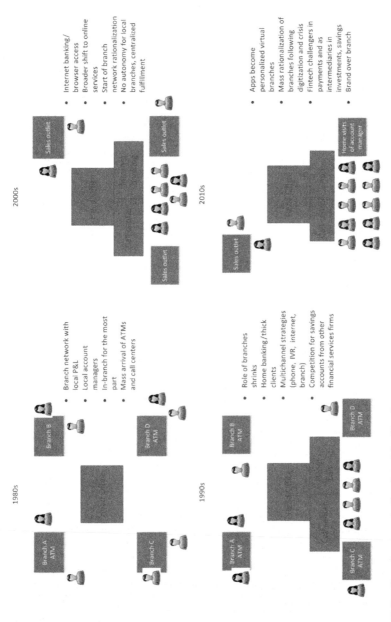

2000s

- Internet banking/ browser access
- Broader shift to online services
- Start of branch network rationalization
- No autonomy for local branches, centralized fulfillment

Sales outlet

Sales outlet

Sales outlet

2010s

- Apps become personalized virtual branches
- Mass rationalization of branches following digitization and crisis
- Fintech challengers in payments and as intermediaries in investments, savings
- Brand over branch

Home visits of account manager

Sales outlet

1980s

- Branch network with local P&L
- Local account managers
- In-branch for the most part
- Mass arrival of ATMs and call centers

Branch B

Branch D ATM

1990s

- Role of branches shrinks
- Home banking/thick clients
- Multichannel strategies (phone, IVR, internet, branch)
- Competition for savings accounts from other financial services firms

Branch B ATM

Branch D ATM

Branch A ATM

Branch C

Branch A ATM

Branch C ATM

FIGURE 3.8   Changing organizational models of banks.

determined by the value of the brand and the quality of the online experience. A good supply chain, information aggregation, and insightful attractive visualization in a customer portal are increasingly determining competitive differentiation.

Many start-ups compete with banks and asset managers. One of the more dramatic areas of fintech development has been in payment services. The move of large parts of the retail sector to the internet has boosted not only parcel delivery services but also payment services for online merchants that address the problem of payment versus delivery (this includes companies, such as Adyen and Klarna). Other new companies have positioned themselves as transparent, trustworthy "no frills" banks or as new-style wealth management companies that offer transparency, low cost, and a superior online experience. (A well-known example would be Betterment on www.betterment.com. Often these firms make use of automatic asset allocation based on customer risk appetites, savings objectives, and investment horizon. This is called robo-advising.)

To some extent investment managers and especially banks are protected by high regulatory walls and their scale. On the other hand, the drag of their legacy infrastructure and increasing complexity of regulatory reporting create opportunities for new firms. The poor reputation of many financial services firms helps as well.

Any online app or platform of a financial services firm should include the following:

- access from any device to a consolidated view of all of a customer's business with the service provider (one view showing savings, current account of all family members, mortgage or other loans, brokerage account and insurance products);
- very easy access to main functions, such as transfer funds and check account balance;
- statistics to analyze and benchmark account balances and transfers as well as portfolio performance;
- customer decision support, clear information, FAQ, and option to go to personal interaction via call, chat, or face to face when faced with investment decisions or when offered additional services;
- options to opt in or out of add-on services, such as location-based data;
- feedback mechanisms and reviews from users;
- reassurance on privacy and security (because a client accesses its information from a host of devices and via different networks, secure authentication is ever more important; information on privacy and security policies and where a client's data get stored should be provided).

New apps or portals are easily launched and provide opportunities for existing providers to compete in markets where they have not competed before. Clients shop around more easily for savings, insurance, or investment products—the checking account that is the core of a banking relationship is stickier. However, some countries have introduced legislation to make bank numbers portable facilitating the switching and increasing competition (see https://www.edgeverve.com/finacle/resources/thought-papers/Documents/bank-account-number-portability.pdf for more information).

3.5 DATA MANAGEMENT FROM THE REGULATORY REPORTING PERSPECTIVE

3.5.1 Regulatory Themes

One of the largest growth industries coming out of the financial crisis has been that of compliance and regulation (see 2016 cost of compliance report, Thomson Reuters, on https://risk.thomsonreuters.com/en/resources/special-report/cost-compliance-2016.html). The policy objectives of regulation are very diverse and include the following:

- strengthening of capital buffers of banks through sufficient reserves and specific attention to additional risk categories, such as liquidity;
- reducing systemic risk in OTC markets through promoting central clearing of OTC derivatives and requiring increased pre- and posttrade transparency;
- living wills to make sure banks have an orderly wind-down plan;
- cultural change by changing procedures, increasing regulation to prevent market abuse, manipulation, tax avoidance, and conflicts of interest;
- prudential oversight and investor protection;
- global harmonization through common data standards for identification and classification of counterparties and financial products.

3.5.2 Regulatory Ecosystem

Regulation has been a delicate process fraught with concerns about making the environment for financial services firms less competitive. Two broad approaches were common:

- Rules-based regulation. This laid down specific rules on banks as to how they should behave and what was allowed. This made requirements clear but also fostered a box-ticking mentality.
- Principles-based regulation. This approach phrases requirements in general terms leaving the details to the banks. In practice it made for "light touch" regulation, which meant that it was always possible to interpret broad principles in a manner that was agreeable to the banks.

Regulation takes place at different levels:

- Global coordination at G20 level. Specific regulatory objectives were kicked off with the G20 summits that closely followed the crisis.
- Global coordination through the Basel Committee of Bank Supervisors. This is a policy-making group of bank supervisors that addresses different

regulatory themes. It focuses mostly on bank solvency. It is not a direct regulator—instead its policy statements are supposed to end up in national law (see www.bcbs.org).

- Regional decision making. This is done at the European Union level that creates either *directives* or *regulations*. Regulations are directly binding on all member states, whereas directives need to be transposed into the national law of each member state. The European Union also has securities, insurance, and bank regulatory authorities (these are ESMA, EIOPA, and EBA for the securities, insurance/pension funds, and banking industries, respectively) that act as coordinators but that also have their own regulatory tasks. In addition to this, direct bank supervision of the larger banks in each of the Eurozone member states moved to the European Central Bank. The European Union typically uses a *passporting* system, which means that companies or services registered in one EU state can be provided throughout all EU member states.
- National regulators. These are often split into *microprudential* (behavioral regulation of ethics, standards, aimed at investor protection, often performed by a securities regulator) and *macroprudential* (monitoring the solvency and liquidity of banks and insurers, often performed by the central bank).
- Self-regulation. This is regulation through self-policing where an industry association is tasked to perform certain regulatory tasks (see, e.g., FINRA in the United States on https://www.finra.org/).

In the United States, regulation increasingly takes place at the Federal level. Following the crisis, the silo-based regulatory framework with tasks split over the FDIC, OCC, SEC, CFTC, Federal Reserve Board, and the (now defunct) Office of Thrift Supervision was seen as a problem. The Office of Financial Research (OFR) has been put into place inside the US Treasury to act as a data service bureau for the different agencies. Regulators themselves need to improve their data management capabilities if they are to act on the data they demand from the industry.

3.5.3 Diversification of Regulatory Attention

The evolution of banking solvency regulation is that of a continuous widening of the set of risks captured. The first Basel Accord introduced minimum capital standards for credit risk. After deregulation and the boom of trading that started in the 1980s, market risk was added as a separate category. Following internationalization and rapid growth in the overall scale and complexity of banks, operational risk was added as a separate risk category for which capital requirements needed to be calculated in the 2000s. Post the 2007–09 crisis, short- and intermediate-term liquidity risk metrics were introduced [minimum requirements on the Liquidity Coverage Ratio (LCR) and Net Stable Funding Ratio (NSFR), respectively]. (The LCR states to what extent banks have sufficient liquid assets to cover cash outflows for 30 days. The NSFR states to what extent banks have sufficient stable funding to cover a 1-year period of continued financial stress.) Banks were already split in product and customer silos—the different risk buckets led to separate functions in the risk department and the introduction of enterprise risk management that aggregated these various aspects of risk.

In addition to strengthening banking regulation, regulators also look at a larger part of the financial ecosystem. Specific attention is given to valuation processes in investment managers (AIFMD), solvency requirements for insurance firms (Solvency II), measuring the liabilities against the portfolios of pension funds, and also specific regulation on service providers, such as credit rating agencies (regulated by the SEC in the United States and by ESMA in the European Union; see also https://www.esma.europa.eu/supervision/credit-rating-agencies/supervision and https://www.sec.gov/ocr).

The complexity of banking regulation (Dodd–Frank Act ran to 2300 pages; see http://financialservices.house.gov/dodd-frank/; also, Thomson Reuters counted over 43,000 regulatory alerts in 2015, up from 8704 in 2008; a regulatory alert is a new policy update, document, or news item) derives partially from the diversity in products. It also derives from ongoing lobbying to prevent simple rules and to create discretion for banks as to how they calculate their risk exposures. A technical focus on some of these models has sometimes led to losing track of the bigger picture and policy objectives.

It is not clear what the next risk category will be. What is clear is that every regulatory reaction so far has led to coping strategies of financial services firms. If risk is made explicit and carries a price tag, it has a habit of appearing somewhere else in a different guise. Next areas of regulatory attention could be concentration risk in clearinghouses (given the larger role of central clearing) or a specific focus on cyberattack risk as part of operational risk.

3.5.4   Increased Process Focus and Data Requirements

A more recent development has been regulatory attention to the information supply chain itself—and not only on the risk models. The European insurance regulation Solvency II already contained some conditions on data quality but attention for data quality has been raised by the BCBS239 risk data aggregation principles (see https://www.bis.org/publ/bcbs239.htm) and similar best practices provided by the Federal Reserve Bank, OCC, Bank of England, and the European Central Bank more explicitly (Table 3.2). Think of someone in secondary school doing their mathematics homework. The teacher does not just want to see the answer but is more interested in the thought process that led to this. Similarly, the data behind the models and the process by which the models are fed need to be sound and accessible for regulatory inspection.

From an information technology perspective, the 14 Principles for effective risk data aggregation and risk reporting come down to common sense. One of the main points of the document is principle 2 on Data Architecture & IT Infrastructure: "A bank should establish integrated data taxonomies and architecture

TABLE 3.2 BCBS239 Principles

Category	Principles	Data management implications
Governance and infrastructure	Governance	Transparency|audit|versioning|granular permission
	Infrastructure	Single repository|integrated|HA/DR
Risk data aggregation capabilities	Accuracy and integrity	Normalization|transparency|full audit|automation|reporting
	Completeness	Multisourced|single repository
	Timeliness	Automation|on-demand
	Adaptability	Flexible|powerful data enrichment and extraction capabilities
Risk reporting practices	Accuracy	Automated|schedule controlled|multiformat
	Comprehensiveness	Full universe/original/cleansed|enriched/internal/external
	Clarity and usefulness	Tailored|fully customizable
	Frequency	Automated|scheduled|ad hoc|fast
	Distribution	Flexible|customizable|user-friendly
Supervisory review	Review	Audit reports|ad hoc requests for information via reporting module
	Remedial	All rules documented|scenario/stress testing
	Home/host coop	Easy rules and validation modifications

across the banking group, which includes information on the characteristics of the data (metadata) ...." This has also proved to be one of the thorniest points for banks to comply with judging from the progress report and banks' self-assessment published in December 2015. The BCBS239 principles apply to the list of global systematically important banks (G-SIBs) as of January 1, 2016, and will be rolled out to different national sets of systematically important banks (D-SIBs) by jurisdiction. (See https://www.bis.org/bcbs/gsib/).

Additional systems requirements include:

- Data accuracy and integrity (principle 3) that puts the quality and control processes for risk data on the same footing as that used in financial statements: "Controls surrounding risk data should be as robust as those applicable to accounting data" (Basel Committee on Banking Supervision, 2013, paragraph 36a).
- A common data dictionary: "As a precondition, a bank should have a 'dictionary' of the concepts used, such that data is defined consistently

across an organization" (Basel Committee on Banking Supervision, 2013, paragraph 37).

- Adaptability (Principle 6), which is the ability to browse the data and satisfy ad hoc requests with quick turnaround: "A bank's risk data aggregation capabilities should be flexible and adaptable to meet ad hoc data requests, as needed, and to assess emerging risks" (Basel Committee on Banking Supervision, 2013, paragraph 48).
- Data quality management requirements as explained in principle 7 on Accuracy. This includes full transparency on the validation rules used in a data management process: "Automated and manual edit and reasonableness checks, including an inventory of the validation rules that are applied to quantitative information" (Basel Committee on Banking Supervision, 2013, paragraph 53b).
- Reporting requirements to group data and to easily report on data on many different dimensions: Data should be available by business line, legal entity, asset type, industry, region, and other grouping (Basel Committee on Banking Supervision, 2013, principle 4, p. 16).

One of the enablers to help firms live up to these principles may be data certification, in other words an attestation by the source of completeness, accuracy, and timeliness (CAT); certainly certified data is becoming more important in an era of more and more raw data.

Apart from a new focus on the process behind risk models, regulators also simply ask for more detail. This includes detailed information on individual loans on both sides of the Atlantic. This includes the AnaCredit project at the ECB; data collection is scheduled to start in 2018 and will cover data on individual loans in the Euro area (see https://www.ecb.europa.eu/stats/money/aggregates/anacredit/html/index.en.html). The objective is to identify, aggregate, compare credit exposures, and detect associated risks on a loan-by-loan basis and granular information on asset classes. In the United States, the FRY 14 Q/M reports require more granular data from bank holding companies on asset classes, and on loan and portfolio level.

Regulators have become more prescriptive in what scenarios and reports they want banks to run. A good example is the stress tests bank need to perform. These stress tests used to be add-ons to the normal risk reporting; now they have become separate exercises with detailed scenarios covering different severity levels on "severe but plausible shocks" provided by the regulators. The point is to force a "what if" way of thinking. Risk management quite rightly is asked to **not** focus on the business as usual.

Stress testing involves subjecting the bank's exposure to extreme events to gauge the effect on the portfolio's value and on the bank's solvency. The data required for stress testing consist of a set of risk factors with their evolution over the time interval of the scenario and specific shocks. The CCAR program in the United States (see http://www.federalreserve.gov/bankinforeg/ccar.htm)

focuses on the top bank holding companies and was introduced in 2011. The DFAST program was introduced in 2014 and includes stress testing for the midsized banks (see https://www.federalreserve.gov/bankinforeg/dfa-stress-tests.htm). Many of these banks were not previously subject to Basel regulation and needed to catch up with their risk scenario infrastructure. In the European Union, the European Banking Authority coordinates the stress tests. The Bank of England has a separate stress test regime (see http://www.bankofengland.co.uk/financialstability/Pages/fpc/stresstest.aspx).

In addition to stress tests supplied by the regulator, banks run their own stress scenarios that depend on their business. Real-life historical events are often taken for stress test scenarios. An alternative could be to check in a historical price database for periods where volatility peaked or where correlations suddenly changed and then subject your current portfolio to those scenarios. Stress tests are applied for all risk domains including market, credit, and liquidity risk. In the case of operational risk there is a special data challenge as most of the available data will reflect small, relatively high-frequency losses. Information on the *tail*—the larger loss events—is very rare and normally consists of highly publicized loss cases.

Global coordination has led to the introduction of the LEI to identify legal entities. The LEI is becoming anchored in new regulations as the coat hanger around which other data elements are organized, for example, relationships between entities, mapping transactions, or exposures to entities. The LEI is being increasingly adopted as a reporting standard in different regulations especially by MiFID 2 to identify all principals and agents in a reported transaction. Regulators trying to get a clear aggregate view of the market have run into similar data quality and identification issues as those that the firms they supervise ran into in their risk and finance department. Consequently, the identification and classification of instruments and counterparties has received more regulatory attention.

3.5.5 Example Basel Regulation: Fundamental Review of the Trading Book

Risk management is a combination of an art and a science. Very often, risk frameworks work very well in normal market conditions; precisely when you do not really need them. One of the new Basel committee policy documents changes the banking market risk measurement and reporting framework (see https://www.bis.org/bcbs/publ/d352.htm). What is new is a focus on the tail end of the returns, in other words a focus not on business as usual but on the unusual, the outliers. Rather than asking the question "what is an upper bound of potential losses in a normal day?", it asks the question "what is the expected loss on a bad day?". In other ways: it asks risk managers to closely inspect the tail end of a return distribution rather than cutting off that tail.

Banks can choose whether to use a sensitivity-based approach or whether to use an internal model approach. To qualify for an internal model approach, banks basically need to demonstrate that they have their house in order from a risk measurement and reporting perspective. For their internal model, banks need to demonstrate that the risk factors they use are modelable.

Eligibility criteria are introduced to enhance data quality and determine whether a risk factor can be used within an internal model. To be eligible data must be:

- continuously available historical observations: 10 years' worth of history;
- of sufficient frequency;
- of high quality;
- supported by real prices (not proxies) at which institutions have conducted a transaction;
- subject to established processes and controls.

There will be an incremental capital charge for nonmodelable risk factors— using a stress test based on the worst case over a 10-year return. The Fundamental Review of the Trading Book (FRTB) increases the historical market data and transaction data requirements. FRTB thus poses specific data management requirements including the integration of transactional with master data, the ability to track and report on long histories for market data and multisource integration and the need for improved data quality management.

3.5.6 Example: Higher Standards in Valuation

Valuation concerns date back to the late 1990s when the increased use of derivatives and several high-profile corporate scandals led to Sarbanes–Oxley and new accounting standards.

There has been an increased discrepancy during the 1990s between the financial risks taken and the information about these risks that was covered in the financial statements. Exposure to derivative products, such as swaps was kept off-balance sheet. This meant that the value of financial statements for investors as a means to understand the health of the company deteriorated. Financial reporting scandals led to an increased emphasis on control. This regulation made the controls over financial reporting even more important. There has also been a move to fair value accounting. FASB 157 refers to fair value measurement and defines a framework to measure it and it implies a need for more transparent securities pricing. (See http://www.fasb.org/summary/stsum157.shtml).

The financial reporting processes of most organizations are driven by IT systems.. Information Technology plays a vital role in internal control: "The nature and characteristics of a company's use of information technology in its information system affect the company's internal control over financial reporting" (for

PCAOB's Auditing Standard 2, see www.pcaobus.org/Standards/Standards_and_Related_Rules/Auditing_Standard_No.2.aspx).

For a bank, good-quality market data is a necessary precondition for accurate financial statements. The Sarbanes–Oxley legislation in the United States for publicly listed companies has given internal controls even more prominence than they already had. This covers controls on the overall process that leads to financial statements and emphasizes separation of duties, clarity on who changed what, when, and why. Sarbanes–Oxley is more *qualitative* (i.e., you have to have and be able to prove a solid process) rather than *quantitative* regulation. One of the key sections of Sarbanes–Oxley is paragraph 404 ("404: *Management's Reports on Internal Control Over Financial Reporting and Certification of Disclosure in Exchange Act Periodic Reports*"), which addresses internal controls on financial reporting:

- Sarbanes–Oxley comes down to a requirement to be able to back up all the published numbers, and by implication also all price, curve, and reference data items that can have an effect on stated earnings or forecasts. This underlines the importance of audit, security, and accuracy of information.
- Regulators and auditors will review how secure, accurate, and auditable the interfaces and the data infrastructure are.

Whereas the earlier valuation concerns came from corporate scandals, more recent valuation regulation specifically addresses investment managers and banks. AIFMD (see http://ec.europa.eu/finance/investment/alternative_investments/index_en.htm) contains provisions for an independent valuation function separate from the portfolio manager as well as for requirements on the custody function following the Madoff scandal where assets turned out to be nonexistent. For the banks, different valuation adjustments are put in place to take provisions against credit losses. Prudent Valuation refers to the use of a representative set of market data sources. More conservative valuation standards for the banking book taking into account different scenarios of credit impairment are also reflected in accounting standards, in particular IFRS9 (see http://www.ifrs.org/current-projects/iasb-projects/financial-instruments-a-replacement-of-ias-39-financial-instruments-recognitio/Pages/financial-instruments-replacement-of-ias-39.aspx).

3.5.7 Summary

In Table 3.3 we discuss several specific examples of new regulation.

Financial firms are in the business of managing risks. The role of regulation is to make sure they report on these risks to their investors, clients, and counterparties and regulators. Regulation will never completely prevent bankruptcies or misspelling scandals—risk has a habit of going undercover for a while and then resurfacing in a new place. More is demanded from firms both in terms of the amount of data they report and on their internal processes.

TABLE 3.3 Regulatory Themes

Regulatory theme	Specific regulation	Process impact/ demands	(Meta) data demands
Solvency of banks	Solvency II, Basel III, Fundamental Review of the Trading Book	New way to measure market risk, tracking liquidity metrics (NSFR and LCR)	Liquidity classification information, increased historical data and real prices for market risk
Investor protection/ trade execution	MiFID2/MiFIR	Pre- and posttrade transparency obligations	Trading venues to report their product master data, systematic internalizers (firms with volume above a market share threshold) to publish two-way quotes, obligation to report all trades
KYC/AML, preventing terrorist financing	Numerous including Patriot Act	Stringent customer onboarding, client identification, and classification	Screening against customer/entity lists, documentation, transaction screening
Raising tax revenue	FATCA (United States)	Customer onboarding, ascertaining tax liabilities	FATCA flag, client classification
Accurate valuations	Changes in accounting standards (IFRS9), AIFMD, PruVal	Independent valuation function (AIFMD), conservatism in valuing banking book assets (IFRS9), adjustments	Valuation adjustments quantification, independent sources of market data, use of real prices
Market transparency	Dodd–Frank, EMIR	Posttrade reporting for derivatives	Classification into instrument categories (ISDA product taxonomy), use of LEI
Accuracy of risk reports	BCBS239	Fourteen principles for sound risk data aggregation	Higher demands on accuracy, audit, and data lineage requirements

LCR, Liquidity Coverage Ratio; *LEI*, Legal Entity Identifier; *NSFR*, Net Stable Funding Ratio.

3.6   BUSINESS DATA ARCHITECTURE

The point of the information sourcing and curation process outlined earlier is to supply business users and information systems with quality information in time. Because automation in financial services firms has taken place along the lines of product, departmental, or customer silos, it has put a high burden on any

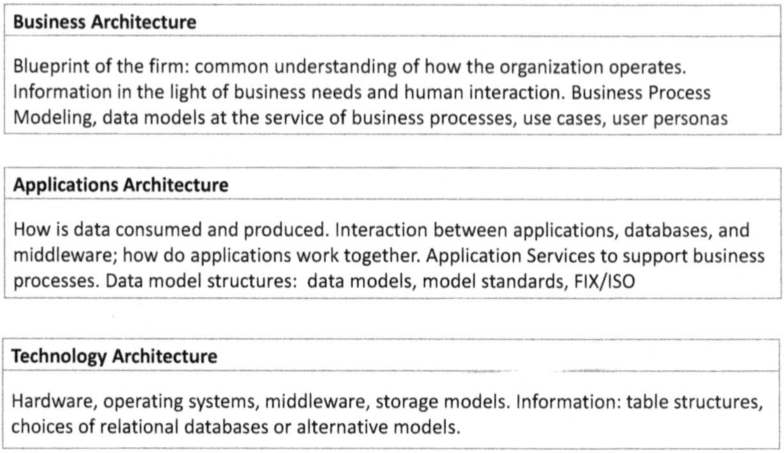

Business Architecture
Blueprint of the firm: common understanding of how the organization operates. Information in the light of business needs and human interaction. Business Process Modeling, data models at the service of business processes, use cases, user personas

Applications Architecture
How is data consumed and produced. Interaction between applications, databases, and middleware; how do applications work together. Application Services to support business processes. Data model structures: data models, model standards, FIX/ISO

Technology Architecture
Hardware, operating systems, middleware, storage models. Information: table structures, choices of relational databases or alternative models.

FIGURE 3.9   Enterprise architecture view.

process that requires an aggregate view. Consequently, it has been costly to aggregate and reconcile different product, market, and client information sets for risk and financial reporting. Similarly, it is costly to create aggregate customer overviews that show different savings, checking, and investment products for a client in time.

To abstract both from individual data sources and from individual application, a business data architecture is needed as the upper layer of an enterprise architecture (Fig. 3.9). (For best practices in information architecture see the TOGAF standard on http://www.opengroup.org/subjectareas/enterprise/togaf).

- Business Architecture is the upper layer of the Enterprise Architecture. This is about modeling the business toward strategic targets and focuses on the optimal delivery of services through business processes that are supported by relevant business roles and information model structures.
- Application Architecture is the middle layer of the Enterprise Architecture. This is based on the Business Architecture and focuses on application services to support the business processes. These application services are provided through application components and functions and adequate data model structures.
- Technology Architecture is the lower layer of the Enterprise Architecture. This focuses on the most suitable technology platforms (hardware, Operating Systems, middleware, and Databases) to support the Application Architecture. Important dimensions looked at Technology Architecture are infrastructure running costs, production stability, scalability, resilience, and security.

The three layers are complementary and strive to a common goal. All three layers have to work together to build an architectural landscape and reap its benefits to the maximum extent.

For most financial services firms the world is changing rapidly: their service offerings have to change and the service level to their clients has to be raised. This in turn requires important business and technology transformation from a process but also from an information management point of view. It is important to conduct this transformation in an aligned manner and with a forward-looking view taking into consideration future strategic aspirations, while achieving delivery agility and operational efficiency.

Business architecture is expected to bridge the gap between the company's business strategy and its successful execution through the achievement of

- a focused and aligned strategy;
- improved decision making;
- increased operational efficiency and capacity from growth;
- agility in business and IT execution.

Thinking in terms of these layers will help to get the most out of the data sourcing and curation process and to provide users with a sufficiently rich and flexible information foundation. We need to abstract from specific business processes and look at an overall enabling business data architecture that services different internal perspectives as well as regulatory and client reporting viewpoints.

3.7 CONCLUSIONS

In this chapter we have looked at the typical information supply chain and curation process, discussing the different steps that take place before data is acted upon. We have discussed the processes by which new financial products are created and maintained. The sourcing, integrating, and vetting of information feeds into a business data architecture: a set of information models that underpins the core business processes with which we started this chapter. A sufficiently rich and flexible business data architecture is a precondition to satisfy both the increased regulatory reporting requirements and customer demands.

Access to information is a critical basis of competition between the various categories of players within the financial services industry. If you know your risk measures and your portfolio attribution, you know what you are doing. If you know your clients and the effects of your actions better than the neighbors, you can also more accurately price a transaction. Regulation on solvency, on investor protection, on sound processes, and on financial reporting promote sound architecture as well. Ultimately, however, more insight into the products and industry participants that you deal with is not a regulatory matter: it provides you with a foundation on which to do subsequent business.

Optimizing the information supply chain across products and processes is what information and process management is about and increasingly determines competitive advantage. An important question is whether the information supply chain of the instrument life cycle *facilitates* or *impairs* successful processing.

Pressure has been put on traditional sourcing and curation processes to accommodate an increasingly diverse set of information sources and disparate downstream information needs. Without a sufficiently flexible and inclusive data collection and integration process, the aggregate picture will be incomplete and end users will leapfrog the "official" process and serve themselves.

Because of the fragmentation in financial products and processes, the content and software application vendor community is growing. In an ideal world, high-quality data on instruments, counterparties, and prices would be available instantly and as easily as utilities, such as power, natural gas, and water. Risk, finance, and especially regulatory reporting sit at the end of the processing chain and for these processes the information integration problem is most acute. Increased regulatory reporting requirements are beyond the data integration and postsilo cleanup powers of risk and finance and have repercussions on processes upstream.

From a client interaction perspective, understanding your user's needs not just in data terms (what they want) but just as importantly in metadata terms (where, how do they need it, how fast, what quality checks) is critical to keep them. Consumers from financial services, whether private individuals, corporates, or other financial services, are increasingly fickle. With lower switching costs, easier means of price discovery, and new concerns on security and privacy, these customer challenges can be added to the business and regulatory challenges. In the next chapter we will look at these challenges more closely.

REFERENCES

Basel Committee on Banking Supervision, 2013. Principles for Effective Risk Data Aggregation and Reporting. Available from: <http://www.bis.org/publ/bcbs239.pdf>.

Faulkner, M.C., 2008. An introduction to securities lending. In: Fabozzi, F. (Ed.), Handbook of Finance. Wiley, Hoboken, NJ, Available from: <http://www.bba.org.uk/bba/jsp/polopoly.jsp?d=130&a=3311>.

King, B., 2012. Bank 3. 0. Wiley, Hoboken, NJ.

McKenna, K., Northey, J., Nichols, B., 2014. Financial interchange standards. Handbook of Financial Data and Risk Information. Cambridge University Press, Cambridge.

Chapter 4

Challenges and Trends in the Financial Data Management Agenda

Chapter Outline

4.1 INTRODUCTION

In Chapters 1–3 we discussed financial data and the role of data management in the main financial services business processes. Information management advantages lead to competitive advantages. All financial business processes have in common that they are very data intensive—and are becoming increasingly data intensive. We saw that a financial institution makes its money through a handful of different activities, all of which face increasing competitive pressures:

A Primer in Financial Data Management. http://dx.doi.org/10.1016/B978-0-12-809776-2.00004-1
Copyright © 2017 Elsevier Ltd. All rights reserved.

- Asset and liability management. The classic banking business of borrowing and lending money, of making money on the interest margin, and of managing the exposure to various time periods of interest. This includes financing areas including mortgages, export finance, and trade finance. This is increasingly difficult in a low or even negative interest rate regime.
- Portfolio management. The business of investing third party assets, usually against a fee that is a percentage of the size of the portfolio reflecting the investment complexity plus often a performance fee. Sometimes (part of the) fees kick in only after a high-watermark or minimally agreed performance is reached. Global assets under management reached $74 trillion at the end of 2014 (Source: BCG; see https://www.bcgperspectives.com/content/articles/financial-institutions-global-asset-management-2015-sparking-growth-through-go-to-market-strategy/). Total profits of asset management industry are at $102 billion.
- Asset services. These include services offered on assets posttrade, such as custody, tax services, triparty collateral, and securities lending.
- Sales and trading. The selling and trading of financial products at the institution's own risk. Proprietary trading for banks has declined following the crisis when the capital markets industry stopped growing. Following a number of misselling scandals previously fertile markets (local government, SMEs) are all but closed off. Many traders have moved to hedge funds where there are less restrictions. The global derivatives market contracted in 2015 to reach $553 trillion at the end of June 2015. The gross market value of outstanding derivatives contract declined to $15.5 trillion at the end of June 2015 (Source: BIS; see http://www.bis.org/publ/otc_hy1511.pdf).
- Investment banking. These are services around wholesale financing via the origination of new financial products (bonds and equities) as well as advisory work surrounding mergers and acquisitions.

In Chapters 1–3 we outlined the topic of data management, the different categories of financial data, and the role it plays in feeding business processes. This chapter will look at the challenges for data management and for data management practitioners. How data intensive are each of these revenue-generating activities, and where are the pain points in scaling, in being more effective in becoming more productive? Where are the pockets for innovations and improvements?

Estimates on spend on IT and operations in capital markets range between $100 and $150 billion and on top of that $100 billion total posttrade and securities services fees (see Euroclear and Oliver Wyman, 2016). What is changing in terms of data availability, collection, and integration capabilities and what are demands from business, customers, and regulators? In short, what is the agenda of data managers?

In this chapter we discuss changing business, customer, and regulatory demands in the light of supply chain and technological developments. In Chapters 5–7 we will look at best practices in technology, quality management, and organization, respectively, to address this agenda.

4.2   CHANGING BUSINESS DEMANDS

Financial services is changing in different ways, in particular around the configurations of techniques and services and more variations on what is done in-house and what services are outsourced. Often, a data management/IT organization can act as a brake on integration and growth. Following the 2007–09 crisis there has been a lot of consolidation in the industry. In most firms IT is a collection of concatenated and poorly integrated systems (where the integration often is effected by Excel sheets and lots of manual reconciliation processes) rather than an effectively executed master blueprint. Infrastructure sprawl is the word: a mosaic of "purpose-built" systems that have difficulty understanding each other. Incompatible IT systems have killed the economics of many mergers.

Poor data management practices can be a major brake on integration and can even completely jeopardize a merger of companies—preventing synergies, such as the creation of common customer databases, and joint services. When adding master data systems on products and customers, in fact the result can be worse than the sum of its parts. Adding new infrastructure through acquisition is just a matter of more data from more silos in need of aggregation. Unifying data from newly acquired systems is no different than integrating any and all enterprise operational systems (which is often a reason why M&A fail; see, e.g., http://www.bain.com/bainweb/PDFs/cms/Public/SL_Getting_price_right.pdf).

As with many design decisions, when implementing or designing systems, it pays off to think ahead and to consider future scenarios of integration and decommissioning. How easily can data be moved out? How easily can additional data be put in? Can we prevent proprietary formats in communication with the platform? What if the volume doubles, triples, or goes up by a factor of 10? Every piece of IT should in fact have a "living will"—an understanding of what it would mean to decommission it. Use case and schema-agnostic database platforms that allow data and metadata to be stored and queried abstract from firm- or department-specific applications can alleviate this.

4.2.1   Key Information Metrics

One of the wake-up calls for information management system was at the height of the crisis in 2008 when it proved very difficult to get accurate information on the exposure to Lehman due to its convoluted legal structure consisting of roughly 2800 different entities sitting below Lehman Brothers Inc. To address this, regulators have asked for more detailed information more frequently and firms are upgrading their reporting capabilities.

However, to avoid drowning in information institutions and regulators have to summarize, to complete, and to transform information continuously as part of the final stages of the internal information supply chain.

The ability to quickly come to meaningful KPIs, shortcuts, and information statistics that convey key properties of the complete data set is ever more valuable. These metrics include

- Metrics on cost that are used by investors to compare offerings. An example is the Total Expense Ratio (TER) of mutual funds.
- Metrics on risk categories defined by regulators to get a quick view of the solvency of a bank, such as the Tier 1 Common Equity and more specific risk categories, such as Expected Shortfall on Traded Risk or Liquidity Coverage Ratio (LCR) and Net Stable Funding Ratio (NSFR) for liquidity risk specifically.
- Metrics on return and risk/return trade-offs, used by investors to pick investments that suit their risk appetite as well as to compare different offerings. Examples are the *Jensen*, *Sharpe*, and *Treynor* ratios, which express different risk measures.

Some measures are used within firms to allocate capital and to be able to compare the efficiency of different departments in using capital to produce returns. Other measures are *risk scales* defined by regulators specifically to help retail investors better understand the risk of the different products on offer.

An ever larger arsenal of KPIs, performance, and cost metrics is thought up by regulators and management consultants. Metrics provide valuable views into the organization and can be extremely useful to zoom in on specific quality aspects of a service, a product line or a firm. They are critical for business management as well as for clients, investors, and regulators to compare offerings and firms and judge whether a minimum value is met.

4.2.2   The Changing Role of Sourcing Departments: The Era of Creative Sourcing

Financial services firms have a growing set of options to configure their organizations. The number of data and software products has gone up and so has the number of deployment options (implemented on-site, hosted externally, combined with IT operations, or combined with business operations). The result of this breakup of the supply chain is often more dispersed operations spread over the world with front office, operations, and IT based in different continents via a "follow the sun" model where tasks are handed over by regions to make for a longer effective business day. This poses management challenges to effectively organize work and handover tasks.

Combined with that, data collection is shifting upstream and more data is captured at points earlier in the processing cycle and injected into operations departments and services in different points of the supply chain. As an example take business entity data. Traditionally a lot of this information was collected directly from the client and business relationships. Due to cost pressures this function is increasingly outsourced to data providers for the basic data and also for the process of onboarding and getting more substantial documentation to KYC-shared services or utilities. A much-cited trend is "everything as a service" (EaaS) that started with Software as a Service where a monthly fee for a hosted software solution replaced an upfront license and implementation costs. This spread to infrastructure as a service (IaaS), platform as a service (PaaS), data as a service (DaaS), and so on (in short: EaaS). From a sourcing perspective this means a move from upfront capital commitments to fund everything out of an operating budget.

Because of increased pre- and posttrade transparency, there is more data in the public domain. This means that the line between producers and consumers is blurring as firms need to report product master and transactional data. It also means that firms can have a more creative sourcing model. We will address sourcing models further in Chapter 7. Technical progress and new service providers and deployment models mean that many different combinations of new technology and business models are feasible. Consequently there is a larger role for sourcing departments in figuring out the optimal combination.

An important value add of an IT department used to be the linking and integration of internal and third party–deployed solutions. This value add is now shifting to sourcing departments to link deployed software, hosted solutions, services, and near-shore/offshore/outsourced operations with internal (Fig. 4.1).

If not done by business managers and system designers, it is up to the sourcing department to have future integration in mind. A key role for sourcing departments is to avoid stickiness and the lock-in to specific data formats and services.

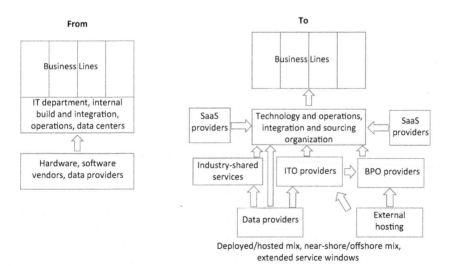

FIGURE 4.1   Changing configuration of IT services.

4.3   CHANGING CUSTOMER DEMANDS

4.3.1   Data Integration Demands

Used to the world of apps and social media, customers have raised the bar when it comes to the accessibility of information and the ease of transacting. Business is done anytime and anywhere and, as for any other information, financial information has "to take flight."

Customers demand friendly apps that provide value add in doing analysis and comparison on their financial products. The financial industry is influenced by the retail industry: the need for no frills products and with swapping/switching options much like a retail returns policy.

The various data categories discussed earlier are often serviced by separate content providers through separate information products. This has been mirrored in the information architecture of financial institutions that typically display a similar siloed implementation. Yet the value lies in combining these data sources and showing interdependencies between different content types. To calculate an enterprise risk figure, market data and correlations on all held instruments are required plus the aggregated holdings. For customer views, the key point is to provide a 360-degree view of all their business.

For retail clients, joined-up data is the imperative and an organization needs to speak with one voice. Integration demands come down to:

- insight into a complete transaction history with allocation over spend categories and with reporting add-ons and trend analysis;
- a view of all long-term financial products with their objectives, such as retirement planning or other long-term savings objectives, together with a current investment style and asset allocation explanation (as provided by new-style robo-advisor firms that have automated asset allocation based on client-stated investment horizon, savings objectives, and risk tolerance);
- presentation of all financial products so they can move from analysis of the day-to-day in an expenditure book and analyze the destination of their income over all fixed costs including mortgage and other loans;
- financial and capital planning tools and execute what-if analysis of life events, career changes, and market movements.

This stretches into advisory services on financial and fiscal planning as we go up the wealth bucket ladder. Historically, retail clients have been quite loyal to their financial services providers. As interaction switches to mobile platforms, the loyalty is to the app, not the account manager. Retail client inertia to switch providers may come down, especially with some countries introducing portable banking numbers and requiring switching services.

In the case of wholesale credit assessment you also need to bring together different data types. For any company to which there is exposure you want to

TABLE 4.1 Different Financial Services Activities

Treasury sales	Credit trading
• Foreign exchange/money market sales	• Walgreens Boots Alliance Inc, USD, GBP, JPY bonds or FRNs
• Cash management services	• Credit Default Swaps on Walgreens Boots Alliance Inc
• Transaction processing	
M&A	**Equity**
• Bank loans	• Cash equity trading on Walgreens Boots Alliance Nasdaq shares
• Advisory services to holding or subsidiaries	• Arbitrage against index
	• Equity derivatives

Within one institution, many different activities are undertaken, all of which translate to exposure to the same ultimate entity. Various types of cross-functional content need to be integrated in order to see the aggregate view. *M&A*, mergers and acquisitions.

know the different ratings, you want to know how its common stock is trading at several listings, and you want to know its corporate bond spread and the price of its CDS. For your internal assessment you want your own interactions with that company, you want to know the legal structure, and you want to have the financial statements and the projections/estimates from the analysts. In addition, you want to be kept abreast of relevant news about the company, its executives, and its industry peer group (Table 4.1).

4.3.2   Client Decision Making

The ultimate result of information processing is decisions. The abundance and ubiquity of information and transaction platforms do not necessarily mean that decisions are taken faster; clients also have more opportunity for price discovery and to shop around. They are helped here by regulators both in the capital markets and for retail investment products. Regulators especially in the European Union put in place standard datasheets on financial products so consumers can compare (see, e.g., the requirements on Key Information Documents for packaged retail and insurance-based investment products on http://ec.europa.eu/finance/finservices-retail/investment_products/index_en.htm) and push for pretrade transparency [e.g., Markets in Financial Instruments Directive (MiFID) II] and more execution venues (the Dodd–Frank Act in the United States introduced the Consumer Financial Protection Bureau; see http://www.consumerfinance.gov/about-us/contact-us/).

Regulators and investors want to track the *context* of decisions made for customers so that they can be reconstructed. *Trade reconstruction* requirements imply the need for increased tagging and recording of information from client meeting notes to phone conversations to preserving the complete context that led to a decision.

4.4   CHANGING REGULATORY DEMANDS

It sometimes seems that all financial services are designed to *complicate*, to *create* rather than reduce complexity when it comes to bridging different currencies, different risk types, market and trading venues, investor risk appetites, and different regulatory and fiscal regimes. Different markets work in different ways, investors have different risk/return horizon requirements, and countries put in place different fiscal systems and different regulatory regimes. On top of that, different parts of the financial services industry (banks, asset managers, insurers) are often regulated differently creating regulatory arbitrage where certain transactions between different parts of the industry make economic sense *only because of different regulation.*

The industry exploits and creates information assymetries in to complex products and contracts to achieve or retain an information advantage. However, the postcrisis squeeze demands from firms the cost base of a manufacturing plant with the cost economies of bulk processing. In other words, the cost base needed implies a high degreel of standardization of products and services. The fundamental industry challenge is to still make a return with these additional reporting obligations and a legacy infrastructure that is often ill-suited to cope with new demands and ill-matched to a manufacturing-style standardization and cost base. (With the exception of the sector of "software publishing and internet services," IT spend as a percentage of revenue is highest for Banking and Financial Services at 6.3% against a cross-industry average of 3.3%. Source: IT Key Metrics Data 2014, Gartner. See http://www.softwareadvice.com/resources/construction-buyer-report-2015/.)

Regulators and industry alike have taken steps toward the standardization of product types, toward improving the transparency on where products are traded and how they are marketed.

Another challenge is the sheer *amount* of regulation: how to efficiently put change management programs together that satisfy regulatory demands. In an ideal world, business demands and regulatory demands would align so that two birds are killed with one stone (or budget). Regulatory objectives include:

- Improve prudential oversight.
- Instill a culture of compliance.
- Reduce systemic risk in over-the-counter (OTC) markets.
- Harmonize global framework (for identification, classification, and semantics).
- Increase tax revenues from capital markets activities.

The data management requirements due to regulation can vary as regulation comes in different types. Macroprudential regulators often have a top-down view when the interest is solvency, so they need aggregated information.

Financial stability can look more granularly at statistics. Microprudential, behavioral regulation looks at specific firms and cases, and therefore at microdata, at transaction and individual actor level.

For the financial services industry, regulation is a mixed blessing. It poses a large burden on reporting and operations in exchange for the license to operate the business. On the other hand, it has also served as a protective wall that has to some extent shielded the industry from competition from new areas. The total postcrisis volume of fines for the top 20 firms as of 2015 is $235 billion (see http://www.ibtimes.co.uk/20-global-banks-have-paid-235bn-fines-since-2008-financial-crisis-1502794). Maybe the interesting observation is not so much the *amount* of these fines but the fact that the industry could absorb them.

Increasingly, firms not only need to sign off the data but in order to do that confidently also need to sign off the whole data sourcing and integration process plus any models used. The protective wall of regulation and the corresponding higher start-up costs have meant that financial services is not like retail where technology innovation can happen faster and put long-established, high-street chains out of business.

Regulation has also led to new data creation. Many pieces of regulation require their own instrument or client identifiers or classifiers and/or reporting flags:

- Solvency 2 introduced a new industry taxonomy outside of industry convention: the CIC codes (see list of CIC codes on eiopa.europa.eu).
- The Legal Entity Identifier (LEI) was introduced following G20 discussions to come to a global standard for entity identifiers.
- FATCA led to new reporting flag to designate the tax status of US securities.
- The European Market Infrastructure Regulation (EMIR) led to a marker on reporting obligations (see http://eur-lex.europa.eu/LexUriServ/LexUriServ.do?uri=OJ:L:2012:201:0001:0059:EN:PDF) for derivative transactions.
- Both EU and US regulations led to a broader standard on financial product identification and trade identifiers. The Unique Swap Identifier (USI) uniquely identifies swap transaction throughout the life of a swap. It includes a unique code that identifies the entity creating the USI and a transaction identifier and this approach prevents two different swaps from having the same USI. The Unique Product Identifier (UPI) uniquely identifies a swap's underlying product. This will be an industry standard for products.
- The increased asset-class scope of MiFID II may well cause the ISIN population to double and grow dramatically after that.

The common thread in this discussion is that data combination and cross-linking improves decision making. In the subsequent text we discuss the following example regulatory cases:

- record keeping
- valuation policies

- KYC
- transaction reporting: MiFID II and EMIR

A common element here is increased transparency on the provenance of information but also on the *data lineage*, that is, the origin of data elements used in pricing and reporting. Regulators want firms to lift the covers and be able to look inside the operation.

4.4.1 Example: Record Keeping

New regulation has increased the requirements for record keeping. This includes:

- the keeping of historical records and the context in which transactions came about to be able to reconstruct a trade;
- audit information and the accessibility of it (what changed, when, how, by who);
- time travel requirements: the need to be able to replay the past;
- transparency, traceability, and lineage of information to trace everything back to its source;
- the fact that most record keeping is on transactional information but new risk requirements, such as FRTB also require lengthier histories of master data and keeping 10 years' worth of price histories.

As an example of record-keeping requirements we look at the Dodd–Frank Act and specifically at the rules for swaps from the CFTC (see http://www.cftc.gov/idc/groups/public/@newsroom/documents/file/sdrr_qa.pdf). First of all, in scope of *transactional* data record keeping requirements are the following pieces of information that apply to Swap Dealers (SDs) and to Major Swap Participants (MSPs), such as investment managers:

- Pretrade execution information. This includes all oral (phone or voice mail) and written (email, chat, text message, fax) communication that led to the execution of the trade.
- Trade execution information. This includes all information entered into an order system that is required for the execution of the trade.
- Posttrade execution information. This includes any operational information on the posttrade process including confirmation, termination, novation, amendment, assignment, netting, compression, reconciliation, valuation, margining, and collateralization stipulations.
- Specific data items used to standardize common swap trade information across the industry. These include the Unique Trade Identifier (UTI) and the USI (see http://www2.isda.org/functional-areas/technology-infrastructure/data-and-reporting/identifiers/uti-usi/). The CFTC uses the LEI, USI, UPI, and also the coordinated Universal Time (UTC), which is the international time standard based on International Atomic Time. This is to make sure that all counterparties in a trade will have consistent time capture.

Nontransactional records are associated with the registration documentation of SDs and MSPs. This typically includes organizational charts, decision-making or governance documents, financial records, audit and compliance information, job descriptions, and biographies and resumes of executives.

The length of the retention period of information depends on the record types. For trade-related information, the retention requirement is set to the life of a swap, plus 5 years. An exception to this is oral communication. In this case the retention requirement is 1 year. For nontransactional information, the length of the retention period is set to 5 years from the moment the record was created. The CFTC is very specific in the type of medium to be used for storing of the required records. CFTC rule 1.31 stipulates the use of a Write Once Read Many (WORM)–compliant medium for storing—meaning once the information is written to the storage medium, you should not be able to modify it.

4.4.2   Example: Valuation Policies

Valuation of financial products is the point where good risk management and accounting policies overlap. We can break the data management challenge up into:

- finding sufficient market data to value the product and potential fallbacks when there is not enough market data;
- taking adequate provisions to reflect all costs and risks associated with holding the position.

On finding the market data, first there is the selection of venues and potential sources from which to draw a valuation price. This selection problem due to the fragmentation of liquidity means you have to keep tabs on much more venues to be able to perform selections, such as "three most liquid venues" even assuming you have a uniform and workable definition of liquidity. (The market models of the various venues may also differ.) Some options to look for market data in valuation include:

- Pick the most liquid value; look at the traded volume and the average size of the transactions taking place at that venue. Check periodically if that venue is still the most liquid one.
- Look at the highest and the lowest price point from several venues.
- Look at the bid–ask spread on the best available quote from all venues.
- Look at the last traded price on all venues.
- Use a VWAP measure, either on one venue or on a selection of venues. The selection can be based on the venues on which the institution normally trades or it can be identical to the set of venues agreed with the originator of the trade (as per the best execution policy agreed with the client).
- Take the price from the local exchange or execution venue.
- If the bank is internalizing (or crossing) transactions, take the trader's mid-price verified against one or more external sources.

● Conversion of prices in case of multiple available execution venues that list the product in different currencies. Take up-to-date foreign exchange prices to do this.

The trend is to take as many sources of valuation information into account and to cast a wide net in finding market data sources. The processes or price consolidation functions (selecting the best valuation price from a series of sources based on criteria, such as date, time, venue, volume, and internal policy) listed earlier have to be clearly documented, whether the process is done in-house or offered by a third-party service provider, such as securities services firms and fund administrators. Furthermore, regulation often prescribes that valuation has to be done independently from the portfolio manager (e.g., AIFMD; see http://ec.europa.eu/internal_market/investment/docs/20121219-directive/delegated-act_en.pdf).

When there are insufficient market data sources available, a *proxy* process can be used. This comes down to using the return or price of similar products to value the product when there is no data. *Similar* can mean comparable in terms of type, (sub)sector of industry of the issuer, quality, cash flow structure, and possibly comparable in terms of duration and convexity. These reference instruments will typically not be fully economically equivalent, that is, responding in the same way to the same fluctuations in risk drivers, so a risk is introduced. A proxy policy stipulates what proxies are used, in what way the price of the reference instrument(s) is used to come with a pro forma value for the original instrument, and when the proxy rule is used. Different inclusion criteria with conditions of decreasing strictness can be set (Fig. 4.2).

If proxies are not used, an alternative is to use model prices—via a mathematical model to value the product based on its cash flow profile. When we use observed market data, we call revaluing a position *mark to market*; when not, we call it *mark to model*.

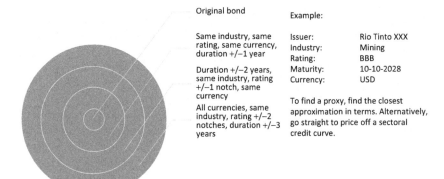

FIGURE 4.2 Bond proxy conditions.

After the financial crisis, valuation policies became more conservative. A position has to be seen in context and what-if analysis on the ecosystem surrounding a counterparty exposure (credit, liquidity, funding) has to take place. Regulators are pushing to fully load the valuation with any impairment cost—to make it "all inclusive." These are adjustments needed to account for the counterparty and funding costs in the risk management of OTC derivatives. This can include different adjustments collectively labeled as XVA and include:

- funding valuation adjustment (FVA);
- debit valuation adjustment (DVA);
- credit valuation adjustment (CVA), the market value of the counterparty credit risk.

Additional valuation adjustments ("AVAs") can also include model risk and are found in prudent valuation policies (see https://www.eba.europa.eu/documents/10180/642449/EBA-RTS-2014-06+RTS+on+Prudent+Valuation.pdf for technical standards on prudent valuation).

Accounting standards, such as IFRS also evolve. IFRS 13 (fair value measurement) defines the concept of "fair value" and provides guidelines for firms on how to conduct valuations, determine fair value, and submit reports. Essentially, fair value is defined as the price that would be received if an asset was sold or if a liability was transferred between market participants on the measurement date. It is the price to exit from the position. IFRS 9 contains the classification and measurement of financial instruments, rules on the impairment of financial assets, and hedge accounting. It stipulates requirements covering the measurement, classification, declassification, and hedge accounting of financial assets and liabilities.

For adequate OTC derivatives processing, extensibility of a data model is important. Many OTC products have been standardized first because of the ISDA master agreements and then by a move to central counterparty (CCP) models. Time-to-market advantages and the higher margin made on products can still justify the higher processing costs and even manual processing. If OTC derivatives are not accurately priced, the institution could suffer from *adverse selection*: counterparties will select the institution as a trading party and continually ask it for quotes with the hope of mispricing.

4.4.3   Example: KYC

KYC refers to the process to verify the identity of your customer. This includes verifying the name of the individual or business against official records. [This can include jurisdictions that are considered not cooperative in fights against money laundering and terrorist financing, such as the Financial Action Task Force (FATF) blacklist. It can also be organizations (see, e.g., http://www.state.gov/j/ct/rls/other/des/123085.htm) or can be secret lists of individuals on "no fly" lists.] This is not only to prevent terrorist financing; the tolerance on tax evasion has gone down and there is more pressure on banks globally to help tax departments collect on money that is owed to them.

KYC is part of the customer onboarding process firms go through to understand and identify clients before they can do any business with them. It is an initial process but also needs to be periodically revisited to ensure information is up to date, complete, and correct in the life cycle of a client. It evolved from verifying the *identity* of the customer to verifying the *integrity* of your customer. It checks that their dealings are above board and that they abide by all local laws in order to combat identify theft, money laundering, terrorist financing, fraud, and tax evasion. It could also include screening against industry sector to abide by certain investment policies (such as Socially Responsible Investing) that do not invest in weapons or alcohol.

After initially onboarding the client, KYC activities include the screening of transactions and money trails and know your customer's customer (KYCC) to look at the counterparties of the client. KYC has been complemented with provisions on Enhanced Due Diligence in the US Patriot Act (see https://www.fincen.gov/fact-sheet-section-312-usa-patriot-act-final-regulation-and-notice-proposed-rulemaking).

Another aspect of KYC is to establish the suitability of the client to trade in certain products. Banks have a duty of care toward their clients and need to make sure their clients understand what they buy. In practice, this means certain products can be offered only following an analysis of the customer. KYC requirements differ by jurisdiction but the general trend is more scrutiny on the onboarding of customers and the tracking of additional information and documentation on accounts.

4.4.4 Example: EU Transaction Reporting—EMIR and MiFID II

Following the G20 meeting in 2009, it was agreed that OTC derivatives contracts should be reported to trade repositories (TRs) as part of the reform of OTC derivatives markets in order to improve transparency, mitigate systemic risk, and protect against market abuse. The aggregation of the data reported across TRs allows authorities to obtain a comprehensive view of the OTC derivatives market and activity.

The EMIR (European Market Infrastructure Regulation) is an EU regulation that aims to improve the transparency of OTC derivatives markets and to reduce the risks associated with these markets (see http://ec.europa.eu/finance/financial-markets/derivatives/index_en.htm). Part of the regulation is to move OTC derivatives to central clearing; part of it is to have posttrade transparency through the reporting of transactions to a TR.

OTC derivatives need to meet certain requirements to be cleared using a CCP. The CCP must be listed in the European Securities and Markets Authority (ESMA) registry and have to be authorized as described in EMIR so that they are recognized across all EU member states. EMIR introduces risk mitigation procedures for bilaterally cleared OTC derivatives and requires all derivatives transactions to be reported to a TR.

Similar to the CFTC rules in Dodd–Frank, EMIR stipulates the use of the LEI and the UTI for reporting to the TR. The UTI is common to both parties to a trade. As the LEI is not yet widely adopted, it must be cross-referenced to proprietary and vendor identifiers used in a firm's counterparty and client data systems. To help with these and other reporting obligations, many firms are centralizing entity data and are creating an entity master with cross-referencing capability so as to accommodate the LEI alongside other entity identifiers.

The UTI poses different problems as EMIR requires all trades to have a UTI, but provides no standard mechanism for generating and communicating the identifier. The result is that UTIs are based on agreements between trading parties. This is currently the scope of a harmonization effort between different regulators (see http://www.bis.org/cpmi/publ/d131.htm on the harmonization of the UTI).

MiFID II is a much broader market reform and amends many existing provisions on business and organizational requirements for providers of investment services, but the most impactful changes are those on the pre- and posttrade transparency regime of EU financial markets.

MiFID II is planned to come into effect in January 2018, and replaces MiFID I (in effect since November 2007). Whereas MiFID I covered only the equity markets, MiFID II details a framework for market data that includes standards, such as ISINs for securities identification, and that will act as a basis for the publication of data to a consolidated tape.

The directive requires a move to faster publishing of posttrade transaction data to local competent authorities and reduces the time delay from 3 to 1 min. This will put new demands on underlying data architecture and means firms need to be able to retrieve supporting reference data from repositories quickly and accurately. Where MiFID II focuses on market infrastructure, MiFIR contains transaction reporting requirements.

The scope of transactions that need to be reporting is growing massively. Under MiFIR, instruments that must be reported include all derivatives admitted to regulated markets, including currently exempt commodity, foreign exchange and interest rate derivatives, all instruments on multilateral trading facilities (MTFs) and organized trading facilities (OTFs), and all instruments that could change the value of instruments trading on any of these venues. The regulation also expands the scope of the data to be reported for each transaction (see https://www.esma.europa.eu/sites/default/files/library/2015/11/2015-esma-1464_annex_i_-_draft_rts_and_its_on_mifid_ii_and_mifir.pdf for the technical standards) to over 60 fields. It has for instance added data fields designed to help spot short-selling traders, and trader and algorithm fields designed to identify the individual or program executing a transaction.

4.5 SUPPLY CHAIN DEVELOPMENTS

Increasingly data is stored "in the cloud." With cloud computing we are referring to internet-based computing using shared processing resources and making these available on demand. There are different versions of cloud computing:

- Public cloud refers to a model whereby the services are rendered over a network that is open for public use.
- Private cloud refers to infrastructure that is operated specifically for a single organization, whether managed internally or by a third-party, and hosted either internally or externally. There may be little difference technically between public and private cloud architecture but security considerations may be different. A community cloud may be a private cloud shared between several parties with similar security and data privacy concerns and based in the same jurisdiction.
- Hybrid cloud is a composition of clouds (these can in turn be private or public cloud) that remain distinct but that are bound together, to offer the benefits of different deployment models catering to different needs.

The total volume of data in the cloud as well as the portion of it is growing (see http://www.cisco.com/c/en/us/solutions/collateral/service-provider/global-cloud-index-gci/Cloud_Index_White_Paper.html). Together with the different sourcing options referred to earlier this means data passes through more hands and is stored in different places. Outsourcing, offshoring, the use of hosting services, and different cloud models all lead to a more convoluted information supply chain. This in turn results in new challenges when it comes to

- Issues in content licensing: content licensing agreements often forbid processing by a third party and do not allow content to leave the premises of the client.
- Confidentiality and data security—data going over networks whether to public or private cloud can be hacked. Some jurisdictions mandate that customer data is stored in the same country as the client for privacy reasons.
- A longer supply chain with more handover points can be more error prone. Additional controls need to be built into it to guarantee quality. We have seen that demands for quality go up because of both business and regulatory reasons. Different quality requirements by different types of business functions within the instrument/trade life cycle complicate the problem and make it more complex to manage SLAs with service providers. Regulators too will raise concerns about who is responsible for what.
- Lower-latency requirements plus increases in product types and volumes put further strain on the information supply chain.
- Banks need to source content and report to regulators in standard ways, using common instrument and entity identifiers and common industry and instrument taxonomies. The formal naming and definition of properties and relationships of entities in a financial services data model is called *an ontology* and puts demands on suppliers too.

Counter to the trend to break up the supply chain and to procure services there is the trend to integrate IT with business functions. Operations and IT have often been combined into technology and operations ("TOPS") organizations and the DevOps movement is a more recent attempt to create a culture and environment where the development and release into production of software can happen very quickly and reliably, with frequent releases.

The challenge for CIOs and CDOs is to keep this supply chain under control and to optimally configure and monitor it. New technologies offer a lot of promise to run operations more efficiently and to make additional data analysis possible. At the same time there are cost pressures on the existing infrastructure. Part of an answer to this is the bimodal IT model, which is the model of creating space for an exploratory IT function while at the same time keeping the business running on the stable traditional platform. Bimodal IT tries to combine running BAU with the least possible risk while creating a laboratory for new technology.

4.6 BIG DATA AND BIG DATA MANAGEMENT

A closer look at a data management infrastructure makes IT often look like bypass surgery—surgical hacks have been applied that were critical to keep the blood flowing. Prevention is always better than cure and the need for frequent bypasses indicates an unhealthy situation.

New technologies can cast a far wider net when it comes to collecting data and putting it through an analysis funnel. They promise to shed light on "dark data"—all data previously hidden from sight of automatic processing. We see only the bits of data that we directly interact with, which are on reports in front of us, when we look up something. Most of the data that could *potentially* influence our decisions, and certainly a lot of the data that is actually underpinning the bits we do see, is lurking below the surface, elusive to our attention. Data is valuable only when it is used and the promise of big data technology is to use alldata generated. We no longer have to take samples; we can use it all. However, enough computing power can beat statistics out of any content and we have to watch that it does not crowd out logic and inferences. 10.000 anecdotes can make a powerful pattern but can also turn out to be no more than background noise not worthy of our attention. Spotting the difference between noise and signal is a critical requirement for analysis tools.

How do we expose the bulk of the iceberg that is under the sea level? How do we lower the sea level to expose the rest of the data? And, importantly, how do we avoid the reverse: that the refined, processed, verified data (the bit of the iceberg above the sea level) is submerged? How do we shed light on the "dark data" prevalent in our enterprises, processes, and business systems? Thanks to more and more tooling to harvest data, it can be used more formally and earlier on in decision-making processes (Fig. 4.3).

Visible data

Changing visibility, expose dark data

Dark data

FIGURE 4.3 Visible and hidden data.

4.7 CONCLUSIONS AND FUTURE OUTLOOK

4.7.1 Commonalities in Client, Business, and Regulatory Demands

A common requirement driven by client, business, as well as regulatory demands is interconnectedness: joined-up data. To paraphrase the law of real estate, the data management agenda can be summed up in three words: connect, connect, and connect. What is new is that this integration of different data sets has a much wider scope: data previously classified as "dark data" that lived outside the realm of automated processing is now also feeding into decision-making processes. The ability to expose and exploit relationships and interdependencies between different data sets can be a major source of competitive advantage.

The supply chains and the configuration of software, data, and service providers are getting more complex while financial services firms look to streamline their operations. Consequently, optimal data logistics—cost-effectively feeding business processes with timely, quality data—is getting more attention. Just-in-time delivery and PAYG consumption models are getting more attention. Compared to just-in-time, there are pros and cons to warehousing data: it is immediately available for consumption and having a local copy means you control it. On the other hand, data gets stale increasingly quickly and the more copies you keep, the greater is the potential for error and confusion.

More attention is also given to the quality of the data collection and integration processes themselves: tracking summary metrics on quality as well as on the speed of supplying the business with data. Master data defines the nodes, objects, and boundary conditions in the firm's data model. The transactional part refers to the business interactions between the actors in the model. Increased traffic volumes, more complex processing and supply chains with storage points everywhere in the world and a larger number of independent entities **acting** on the data complicate the picture.

4.7.2 Toward a New IT Organization Model—Securing Pockets of Innovation

One way for firms to take advantage of new technologies and processes is to embrace the concept of bimodal IT, more specifically, to identify *pockets of innovation* where firms can experiment and try out new models, concepts, and technologies. This should take place on stand-alone processes as a Proof of Concept before rolling them out to the main business functions so as not to take unnecessary risks. How do you spot improvement pockets for data management where you can innovate from within?

Some of the characteristics of these pockets of innovation are listed in Table 4.2.

The question is to what extent firms are willing to take bigger technology bets. Firms may have to spend more on innovation to differentiate as captive

TABLE 4.2 Characteristics of Pockets of Innovation

Characteristic	Explanation
Independent	Should not depend on adoption discretion of third parties. Complete control over the process within the organization
Near-term benefit	Demonstrable benefit within 18 months—should not cross over more than two budget cycles, else political capital will be gone
Not requiring mass adoption	Can work for platforms that can gradually siphon off traffic from a legacy platform. For example, the benefit must be clear when capturing <10% of the volume of an existing process. Could, for example, work in client interaction processes or order fulfillment. Harder where there are natural monopolies, for example, in liquidity pools
Incremental alteration in clearly defined market space	Focus on specific, actionable use cases where a small number of initial participants are required to gain critical mass. Alterations to narrow areas of existing processes or smaller market transformations could work
Clarify costs	Expose current costs and operational challenges to isolate the improvement area and scope of the test case and innovation
Prototype fast and plan for multiple iterations	Learn by doing. Expose areas needing refinement. Success breeds further innovation

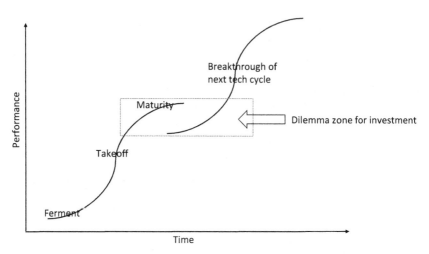

FIGURE 4.4   Technology adoption curve.

markets have disappeared. In many areas of the financial services industry, competition goes up and the playing field is the world.

4.7.3   Concluding Comments

The financial services industry has a poor track record of standardization and adopting new technology standards. It is inherently a conservative industry that has been shielded by protective walls of regulation. In master data standardization and the use of global, unique component identifiers, the industry is still significantly behind other industries.

Also, capital markets is a networked business—often the industry moves either en masse or not at all. Technology innovation will follow the innovator's dilemma adoption curve based on Clayton Christensen's *The Innovator's Dilemma*(Harvard Business Review Press, 1997)(Fig. 4.4).

The inflection point in any innovation will be reached when there is a large pool of people with skills and a large number of peers that have already moved, leading to a sufficiently high comfort factor for decision makers. In the next chapters we will go into more detail on the technological, process, and organizational approaches to securing an optimal data management practice in a financial services firm.

REFERENCE

Euroclear, Oliver Wyman., 2016. Blockchain in capital markets—the prize and the journey. White Paper, p. 20.

Chapter 5

Data Management Tools and Techniques

Chapter Outline

A Primer in Financial Data Management. http://dx.doi.org/10.1016/B978-0-12-809776-2.00005-3
Copyright © 2017 Elsevier Ltd. All rights reserved.

5.1 INTRODUCTION: TECHNOLOGY ENABLERS

The lack of a sound information architecture poses a continuous additional cost and in the end, there aren't many institutions that will be able to continue to afford it. The financial services industry was early to use automation and tools—and is suffering because of it. In this chapter we will discuss different approaches to information technology and provide an overview of the typical components of the IT stack. We will cover the IT responses to the demands sketched in Chapter 4. We start with the types of tools provided and their trade-offs. We then discuss the traditional infrastructure building blocks in the form of the underlying data types and storage options, as well as technology developments.

We compare different choices of storing information and different models of organizing. The information management structures can range from light (files, indexed or not) to much more organized with retrieval and management options (databases) including functions, such as backup, replication, access, change management, and granular data access permissions.

When we move from hardware to operating system, to file system to databases to application software, you get increasingly closer to business processes and individual preferences, until you end up with end-user computing tools, such as Excel. The end goal of IT should be to provide a set of reusable data services ready to use and put at the disposal of the business users. In other words, an easy-to-reach, easy-to-exploit data operating system, self-service data layer for the user. Although it should be self-service, such a layer should include a feedback loop to rank data quality and suggest improvements or complementary data sets. It could also be a direct contribution channel to not only collect feedback on the quality but also directly pool data that originates with the business users in a crowdsourcing model (Fig. 5.1).

Lastly, we have a class of distribution and "extract, transform, and load" (ETL) tools. The need for those arose with decentralization of applications. New challenges for IT infrastructure services include a further decoupling between the point of usage and the point of storage of the data: the cloud, as well as a proliferation of devices through which both staff and clients reach the infrastructure. This brings new security challenges.

5.1.1 Different Levels in an IT Infrastructure

An IT infrastructure can be described using different metrics that define properties at different levels:

- At the hardware level we are interested in computing power and storage speed and amount. We care about the capacity of the network and in non-functional aspects, such as security and backup guarantees. Some of the traditional constraints have mostly gone away in terms of storage and CPU power.

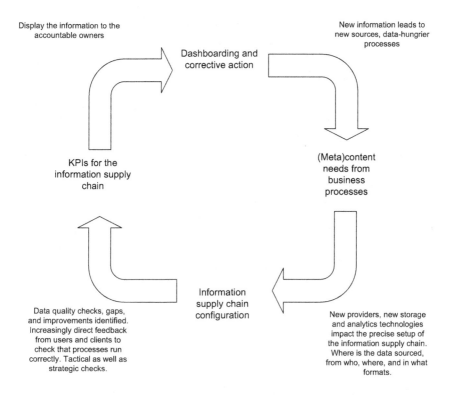

Display the information to the
accountable owners

New information leads to
new sources, data-hungrier
processes

Dashboarding and
corrective action

KPIs for the
information supply
chain

(Meta)content
needs from
business
processes

Data quality checks, gaps,
and improvements identified.
Increasingly direct feedback
from users and clients to
check that processes run
correctly. Tactical as well as
strategic checks.

Information
supply chain
configuration

New providers, new storage
and analytics technologies
impact the precise setup of
the information supply chain.
Where is the data sourced,
from who, where, and in what
formats.

The continuous feedback loop requires an in-depth understanding
of process and of the meaning of the metadata that is tracked

FIGURE 5.1 **Data feedback loop.**

- Looking at the infrastructure software, such as the operating system and databases we are interested in support, security, business continuity, and access to a pool of skills to source this support. We can evaluate required database properties based on the data domain (density of network relationships, frequency of time series data, and general business purpose).
- Key aspects of infrastructure are scalability: both horizontal and vertical. *Horizontally* scalable means that solutions scale well if you add additional machines into your pool of resources. *Vertically* scalable means that solutions scale if you add more power (CPU, RAM) to your existing machine. In databases, horizontal scaling is often based on splitting the data over different nodes; with each node containing part of the data. In vertical scaling all data resides on a single node and scaling is done through adding more cores, that is, spreading the load between the CPU and memory resources of that machine.

- At the application software level, we see the specifics of the business processes, the specifics of the primary or secondary financial services business processes discussed in Chapter 3. For instance, there are applications for order management, portfolio management, core banking, execution management, risk management, trading, and fund administration. They often come with domain-specific data models represented in a commercial DB solution. Important elements to consider here include the rights in the license (to avoid hidden cost or sudden cost jump triggers), support conditions, options to influence the future direction of the software, and getting in change requests when you need them, with escrow and source code access as a last resort.
- Given the number of applications in circulation (For example, Deutsche Bank has amassed over 4400 applications and is not alone in wanting to materially shrink that number; see http://blogs.perficient.com/financialservices/2016/05/03/deutsche-banks-big-it-plans/), good integration capabilities are a prerequisite. Application management vendors should focus on application programming interface (API) management capabilities for third-party best-of-breed tool integration, which ensures that organizations are well equipped with an internal software development system that can easily integrate with any IT or ALM tools needed in a project.
- How a firm integrates different business applications is important. Some business applications will be built internally as some firms have unique business needs not met by any commercial vendor or have a culture of DIY. Desktop applications and end user–developed applications (EUDAs; spreadsheets, local databases) can boost productivity but can turn into a material risk if they start to occupy steps in a business critical production process.
- On the overall IT service management you look at metrics, such as being future proof—have a clear picture of the implications of adding users, adding transaction, or master data volume. How do the costs scale with usage? Can I control the change management process?

Metrics, such as TCO are meant to take all factors into account: the cost of a piece of software not only is the license itself but also includes the hardware, any other software licenses needed as OS and RDBMS, as well as the cost of the staff running it. However, TCO in itself is too narrow. Requirements on the applications and infrastructure are likely to change. On the one hand, do not sacrifice flexibility and extensibility for current TCO improvement. Opportunity cost has to be taken into account but also do not build for the next century. The question comes down to the life span of applications. The answer is the following: longer than you think as applications are easier to build than to decommission. Whereas the life span of a spreadsheet may be a couple of years, applications with their own database last around 10 years and core backend systems around 20 years. This means that flexibility in data integration and extensibility, and abstraction from application specifics in the data model will pay off when the time comes to decommission the software. This element has to be factored into the TCO picture.

5.1.2   A Short Taxonomy of Data Management Tooling

Different tools to manage data include:

- Metadata management tools. These tools are to manage the definitions, the access permissions, the relations, the types, and the permitted values.
- ETL tools. These tools are to format data, to discover structure, and to store it in a database.
- Data discovery tools. These tools scan (large) sets of data to find the relationships and the rules hidden in the data. This can also include data profiling, which is the use of analytical techniques to document the structure, content, and the quality of data.
- Data cleansing tools. These tools find errors but also standardize addresses (zip or post codes, street names, towns, countries) elements, nomenclature, and company or customer types. A common use case is to find all people in the same household from a large dataset.

An IT infrastructure exists to process data. There are different ways to look at data. One useful way of classifying digital data is by three different states. Data can be said to be "at rest," "in motion," and "in use." With this the following is meant:

- *At rest* refers to data used in storage. It includes storage on local drives, on backup, or files on servers in a storage area network (SAN). The data needs to be secured since it often contains sensitive information and is retained for longer periods often for legal reasons.
- *In motion* refers to data that is transported either over an internal network or over the internet. It could be sent via different protocols or email.
- *In use* means data is processed by an application or end users—and will typically be extended or modified.

Data at rest is covered when we discuss data storage options in Section 5.2. From a data in motion perspective, there are essentially two types of supply chain:

- The push model: information comes in from outside and is warehoused in the IT stack, that is, an information inventory is kept independently of direct use by applications.
- The pull model: information is requested when it is needed.

Often there is a mix of these models where commonly used information is kept in internal databases (Fig. 5.2).

Proactively source, push model. "Build to inventory" data
management process.

Data sources pushed into firm

Reactively source, pull model. "Build to order" data management
process.

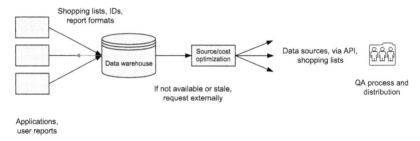

Applications,
user reports

FIGURE 5.2 **Push and pull models for information provision.** *API*, Application programming
interface.

If we look at the "data in use" state, there are many different kinds of applications that will lock data to be processed. We can classify applications by what they do in terms of information status as follows:

- Reporting and reformatting of data—focus is on the ***presentation*** of data.
- Analysis of data—focus is on calculations and drawing inferences, the ***reasoning*** with data.
- Processing—focus can be scripted data entry or on transactional data changes, such as in payments; or trade processing—focus is on ***data operations***.
- A separate category of processing is in making ***master data changes***, for example, in maintaining a customer or product database, onboarding new clients, and entering new products.

If we look at the typical IT stack, we move from hardware up to an operating system, up to applications and databases, financial data models, and end-user tooling (Fig. 5.3).

The lower you go in the stack, the more room there is for standards. Important technology standards and standards organizations include the following:

- The World Wide Web Consortium (W3C) is the international standards organization for the World Wide Web.
- ANSI (America) and ISO standardize languages, such as SQL.

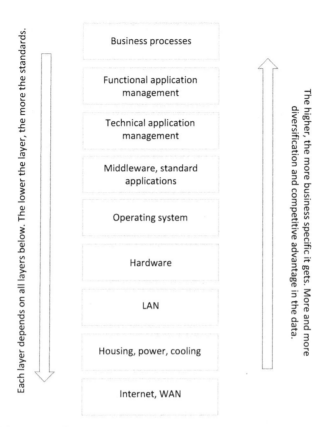

FIGURE 5.3   Typical IT stack layers.

- CMMI is a methodology for the appraisal and scoring of the maturity level of an IT development organization.
- Object Management Group (OMG) is an industry consortium focused on the development of standards for distributed object–oriented systems. It focuses on models and model-based standards.

With drawn-out business processes that span many departments, countries, and applications, it is not only the scheduling and managing of jobs (schedulers) but also managing the transport of the data (middleware), setting up workflows across the processes (BPM), and simply managing who owns the data and who can do what to the data (Data governance). In between core infrastructure (network, OS, RDBMS) and application software, we have the categories of tools shown in Fig. 5.4.

5.1.3   Data Governance Tools

Given the variety of places in which data is at rest, in motion, or in transit, it is important to keep track of data definitions. To get the most value out of data, users in departments from finance and operations to front office are becoming

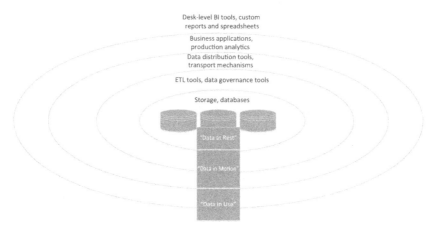

FIGURE 5.4 Categories of tools. *ETL*, Extract, transform, and load.

more involved in information management. This process needs coordination to make sure different users have a common understanding of the data they use. Data governance tools are collaborative tools that capture business and data definitions, critical attributes, define and track relevant and actionable KPIs, and involve business data stewards in the review or approval workflows of these definitions. After this the results are published to the entire organization.

Processes to be supported include data quality management, data sharing agreements, change requests, and managing the collaboration on defining and maintaining data assets across different stakeholders. Workflow processes, policies, and data model definitions can be created, reviewed, approved, and enforced. Metrics on data quality and compliance with usage policies should be tracked to increase the confidence in the data.

Especially in master data, to prevent ambiguity and to ensure a common data language, you need a controlled approach to handling data domain specifications, controlled vocabularies, hierarchical reference data, or mappings between various versions and variations. This will avoid broken business processes and costly integration challenges and will speed up STP and the industrialization of processes. Reference data is a data stewardship problem; we will further discuss this in Chapter 7.

Data governance tooling is one example of collaboration software. Because many users will more frequently touch data or metadata, social media–type tools are spreading to business workflows. Users will nibble on data and tools will help to organize communities of interest, expertise, and stakeholders.

5.1.4 Analytics Tools

Tools for reporting and analysis are collectively called business intelligence ("BI"). Examples include IBM COGNOS, SAP Crystal Reports, and Palantir.

This is an area that is rapidly developing given the growth in available data that analysis tools can act on. Usage can range from simple reporting to large-scale pattern detection. Organizationally, these tools often occupy a slot in between core infrastructure (transaction processing systems, master databases) and end-user desktops (spreadsheets and word processing). The wider use of analytics tools will push up data quality as the ramifications of poor quality will increase. Second, analytics tools will also be directly used to improve the data quality as quick scans and profiling of data will become standard operating procedure on new data entering an organization. There will be stricter controls on data entering an organization. The data stays with us a long time and the average life span of data is going up because of retention requirements.

5.1.5   Data Distribution Methods

Content can be made available and distributed in different ways. They include the following:

- *Fixed delivery times* that are based on scheduled cutoffs of earlier processes.
- *Event based.* Distribution can occur at any time and with irregular intervals and is based on trigger events, for instance, news or an updated term sheet.
- *File Based.* Normally, files have clearly defined structures and sometimes files are specially cut for clients. In that case there is flexibility in selecting which information fields are included and which data is included. This selection can take place through a web browser or through supplying a shopping list file to the vendor.
- *Streaming data.* Here a client retrieves a continuous price stream, direct from exchanges via a direct market access feed, through ticker plant software or through enterprise data providers. This pushing can be done to user's desktops, their email, their smartphone, tablet, and so on. A typical example is a trading terminal displaying real-time quotes.
- *APIs.* In this case, applications at the user site can directly access vendor content through using the vendor's API. The API will outline a number of calls in which information can be retrieved. As data has to be requested and directly ends up in applications, data usage can be more easily measured and controlled.
- *Web Services* make use of a more standard architecture using XML and http. Client applications can invoke services to request or commit data. For the data provider, this has the same advantage as far as keeping tabs on data usage is concerned.
- Browsers can be used to directly consult a vendor's database and terminal products can be purchased to make use of vendor data.

The main characteristics of the different media of providing data are summarized in Fig. 5.5.

The trend for information to go electronic started with the ticker tape which Edison patented in 1869, through the rise of the telegraph and telex, through real-time feeds from Thomson Reuters or Bloomberg, and through the move

	Request/response	Publish/subscribe
Content not curated	Snapshots off streaming data, for example, pulling quotes	Continuous event-based data hot off the press, streaming to message queues for example, newsfeed, tick data push, terminals
Content curated	Self-service-based data products, for example, APIs on data warehouses and data services	Master data made available at certain times of day, for example, end-of-day pushed files and reports at end of production process

FIGURE 5.5 **Different data sourcing characteristics.** *API,* Application programming interface.

from reports first to CD ROMs and then to file- or message-based electronic feeds with the rise of the internet. Storage options continue to develop and cloud storage is now a commodity.

By moving from bulk files to providing more granular API-based access via customer shopping lists, users of content can request specific pieces of information on specific instruments, legal entities, or customers. In this way, data vendors help to eliminate unnecessary data traffic, processing, and storage. On the other hand, through working with shopping lists of specific items, clients also provide exact use case information to data providers every time they make a request. License agreements are increasingly tied or limited to specific use cases. The ideal scenario for a content provider is to know every use case and each moment a piece of data is used in a specific business process at their client.

Indirect vendor access methods include the aggregation of files through an internal quality assurance process before making available the content to users and applications. If files are designed to be database ready, that is, set up for automatic processing by clear structuring, self-describing through column headers or tags, and an unambiguous and complete data dictionary, this will facilitate processing. Information can pass through different curation stages before being stored into data warehouses or operational data stores. We discuss this process in Section 5.2.3.

Direct access methods can be used for content that affects only local decisions or where every microsecond counts. Indirect access methods will slow down the process of making content available but can improve quality and add value by providing an integrated overview of information.

Within an IT infrastructure, there is a trend to open up every application with APIs. To use a metaphor, if applications were people, they used to send each other letters (files) that they had to open, read, digest, and then write back a response, maybe the next day. Some letters got wrongly delivered, lost, or simply misunderstood. Now we are moving to a continuous conversation or chat room model. Applications listen continuously for incoming information and when they get a question, they quickly respond with the requested information.

Once the information is in-house, there are several ways of distributing it. It can be made directly available to end users by pushing through certain events or by having applications directly interface with vendors, or it can go through a quality assurance process. The more applications there are, the larger the data logistics problem. Middleware is the category of tools to rout data internally to where it is needed.

Typical middleware functionality includes messaging (the communication between applications on different platforms), the integration and the workflow (the capture, visualization, and automation of business processes), and orchestration—meaning the coordination of data supply chain processes.

Middleware manages the "data in motion" and is the superglue in an application stack. If we think of financial services as a data manufacturing plant, then middleware is about supply chain management and logistics: the transport and scheduling and making sure data arrives as fresh as possible (just in time) at the point of consumption. Waste in this metaphor means idle time of data processing applications. One of the main tasks of middleware is to optimally allocate data flows to users and applications and to watch over the freshness and thereby usefulness of the data. Specific tools, such as BMC CTRL-M, Autosys, and IBM Tivoli manage IT processes and schedule, track, and analyze activities.

5.1.6 EUDA

The acronym EUDA refers to end user–developed applications, which means any application fully controlled by a desktop user. The most common example is Excel but EUDA also include local databases and scripts that can sit inside Word macros or email programs.

EUDAs are often made for functionality for which enterprise-wide solutions cannot be developed, either because there is no time or because it will not be economical. Local spreadsheets and databases are fine as long as they act as curation tools to prepare data for local use or as an end report. However, they can represent material operational risks since

- The EUDAs are typically created without supervision, security, and backup—typical elements of a normal software development process. EUDA are idiosyncratic individual DIY tools by definition.
- EUDA can overlook compliance guidelines on the use of sensitive data.
- Because they are developed by individuals, they represent a single point of failure.

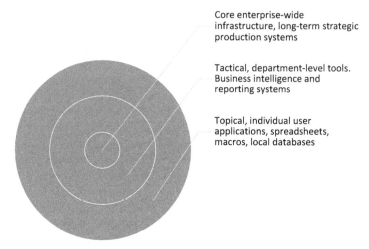

Core enterprise-wide infrastructure, long-term strategic production systems

Tactical, department-level tools. Business intelligence and reporting systems

Topical, individual user applications, spreadsheets, macros, local databases

FIGURE 5.6 **Infrastructure, BI tools, and EUDA.** *BI*, Business intelligence; *EUDA*, end user–developed application.

The governance and control process is the main risk here. Because of the flexibility and time pressures, it is common that spreadsheets creep into critical business processes and are not just used as end points to present data but as data processing tools on whose output another production process directly depends. The support is simply not there—it is common that an error is known to be somewhere in a spreadsheet but that the only person who understands the spreadsheet happens to be on holiday.

To be clear: tools, such as Excel are a blessing for local use and will never go away. Sharing knowledge of important spreadsheets and using version control on them will help. The main risk mitigant though is that of mindset. EUDA should be seen as local data manipulation and as reporting tools—not as production tools. They should always sit at the end of a data flow, never in in the middle of a production day-to-day workflow.

EUDA applications sit close to the user and the analytics tools discussed earlier often occupy a middle slot under the label "BI" in between core production systems and desktop software (Fig. 5.6).

The differences between production systems, BI, and EUDA are summarized in Table 5.1.

5.2 DATA STORAGE MODELS

Information is recorded and expressed in data types. These classify at a low level the categories of data including what the possible values are for that piece of data, the meaning, and what kind of operations can take place on the

TABLE 5.1 Differences Between Production Systems, BI, and EUDA

Production systems	BI	EUDA
Inflexible, solid change management, longer change cycles	Reporting applications	No IT department needed, short change cycle, no version control
Very fast and stable	Separate group in between business and IT	DIY data collection
Back-end infrastructure in data centers	Small servers or local computer	Local computer
Core of IT department	Tolerated by IT department	Invisible to IT department unless there is a problem
Data security	Intermediate data security and quality	High risk of poor data quality and data loss
Formal SLA embedded in support organization, clear service window, and escalation procedures	Often informal support, ad hoc	Users are on their own

BI, business intelligence; *EUDA*, end user–developed application.

data. Examples include a real number, an integer, a string, and a Boolean (a yes/no flag).

What basic operations can act on data depends on what the data type is. Addition would work on reals and integers but not on a Boolean. For a string it would have to be defined more closely (it could, e.g., mean concatenating two strings). Logical operators like OR and AND would work on a Boolean data type but not on other data types.

Combinations of several data elements using these basic data types could be used to come to more complex data types. For example, colors are three integers representing the amounts of red, green, and blue (the RGB scores) and complex numbers can be made out of two reals. Large sets of real numbers can be used to create composite data types often used in financial market data, such as curves, surfaces, and matrices.

The choice of data type is the first level of interpretation done on the data. Using the wrong data type could lead to information loss. For example, treating an interest rate as an integer instead of a real could truncate the part behind the comma. Furthermore, the choice of data type determines what basic operations can be applied. 1234 could be stored as a string or as an integer—if stored as a string, this would open up operations, such as appending it to another string, but preclude addition or subtraction.

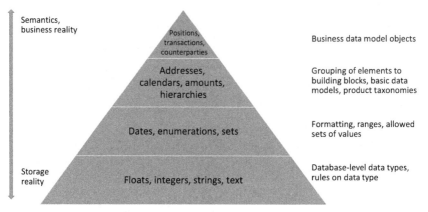

FIGURE 5.7 **From data types to semantics.**

A second level of interpretation takes place through screening of the values of the data in data types, such as *date* and *enumerated*. Date fields can be combined with a calendar, a list of the bank holidays, and weekends to come to a set of business days. In enumerated fields the value of the data type is drawn from a predetermined set of fields, say a set of currency codes, industry codes, country codes, or client types.

If we encode more business rules on the data, we can build meaningful business objects. Apart from the trading date example (calendar + date field), examples include amounts (currency code + a real number) and a payment (amount + date + destination bank account number) (Fig. 5.7).

In this section we discuss different ways of organizing data and corresponding storage technologies. In 5.2.1 we discuss different databases commonly used; we devote section 5.2.2 to NoSQL technologies.

5.2.1 Data Modeling and Databases

What are the methods to record and access data? For data at rest? What is changing in the big data era? Using the data types discussed earlier, the financial data types discussed in Chapter 2 are organized into models to support business processes.

5.2.1.1 Organizing Data

The simplest data model and a widely used approach is that of a star schema. A star schema consists of fact tables that reference dimension tables. The star schema is a special case of the snowflake schema. It is effective for handling simple queries (Fig. 5.8).

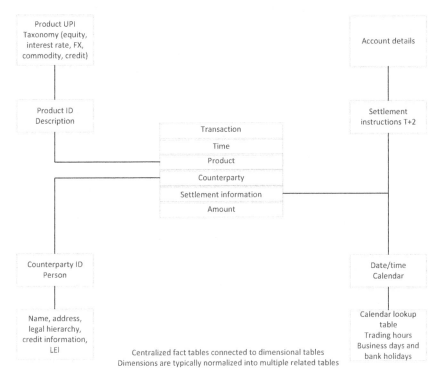

FIGURE 5.8   Star schema example.

5.2.1.2   Relational Databases

The different database technologies differ in the following:

- What do they consider as a logical unit of storage?
- Is data stored on disk (or other storage device) or kept in memory?

The relational database ("RDBMS") originated in the 1970s and the underlying relational model was invented by Codd (1970). Relational databases present data as a collection of tables with the rows being the values belonging to each table and the columns being the attributes of that table. A table could, for example, be all the attributes of a client, a financial transaction, or a product.

Tables have a primary key that uniquely identifies each record. They can have foreign keys as well that link to information in other tables. The term RDBMS often goes hand in hand with SQL. SQL is the syntax of how to interact with the tables to extract data or to update data (Fig. 5.9).

Transactions in relational databases conform to the following ACID principles:

- Atomic: this guarantees that a database transaction is either completely executed or not at all.

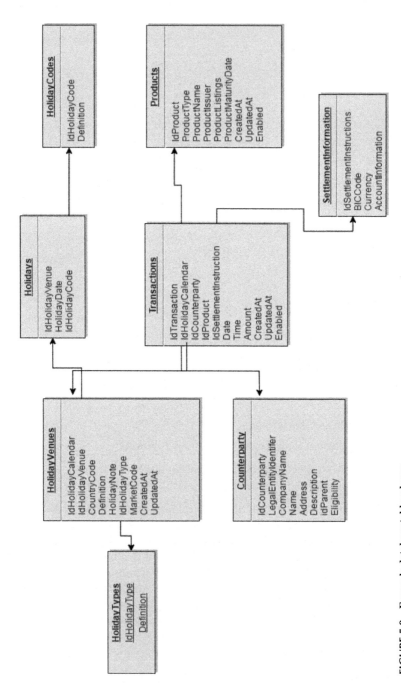

FIGURE 5.9 Example database table schema.

- Consistent: a transaction leads to a new state. In case of an error, the database will fall back to the previous state.
- Isolated: transactions are executed independently, that is, interim results of one transaction do not influence another transaction.
- Durable: a transaction that has completed cannot be reverted.

More recent distributed systems that originate in the social media world (NoSQL) do not conform to the ACID rules as the consistency criterion has been relaxed. An alternative set of database rules has been coined: Basically Available, Soft state, Eventual consistency (BASE). We will discuss this further in the subsequent text.

A key concept in relational data models is that of normalization. This is a procedure to organize the attributes and relations to come to a minimal amount of data redundancy. The idea of normalization is that data is stored only at one place (so you will have to modify an attribute only in one table) and that it will be propagated throughout the whole database using the foreign key relationships.

5.2.1.3   Alternative Databases

Following the adoption of RDBMS in the 1980s, several challenger database technologies appeared. In this section we discuss XML databases, object databases, time series databases, and in-memory databases. How to considerations on choosing ways to organize information include:

- facilitate the efficient loading of data;
- facilitate easy and flexible data manipulation: combining data in different ways across different dimensions and performing calculations;
- facilitate reporting, fast access, and retrieval of data.

These three aspects do not go together, so data stores lean toward one of the three typically at the expense of at least one other aspect (Fig. 5.10).

5.2.1.4   XML Database

Whereas in a relational database the logical unit of storage is a row in a table, in an XML database this is an XML document (which can be small or big). XQuery is the query language rather than SQL. XML documents use tags to express interpretation or meaning and to facilitate the processing of the data by applications. There are a number of reasons to directly specify data in XML or other document formats, such as JSON:

- The XML format is often used for data transport because it lowers the risk of ambiguity at the handover point. Assuming you have agreed upon the tags and the structure, you might as well store directly in XML. The ability of the native XML database minimizes the need for extraction or entry of additional metadata to support searching and navigation.
- Data may need to be exposed or ingested as XML so using XML means staying closer to the process and you weld data at rest and data in use together and avoid double modeling of the data.

Transaction docs:
{transaction ID
Details
{product id
product type
Terms and conditions}
{counte party id
Counterparty details}
{settlement info
Account details}}

Collection of complex documents with nested data forma:s varying record format

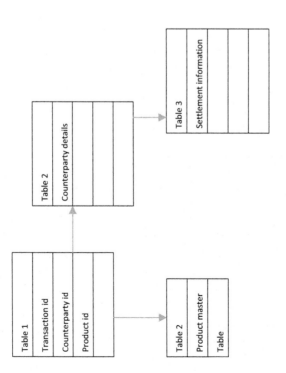

Structured tables with rigid data formats and records

FIGURE 5.10 RDBMS versus document database representation.

- XML is well suited to sparse data with a lot of blanks or to represent nested data.
- XML is more easily human readable, whereas relational tables require expertise and the use of SQL queries to access.

5.2.1.5 Object Database

An object database is a database management system in which data is represented not in tables but in the form of objects. This type of database is closely linked to object-oriented programming. Any object is a combination of both data structured and functions and variables.

An *instance* is a concrete occurrence of something and a *class* is a template for creating objects.

Unlike a relational database, an object database is much more closely integrated with the programming language. An object database stores complex data and also the relationships between data directly, without the mapping to relational rows and columns. This makes object database fit for applications that deal with complex data.

5.2.1.6 Time Series Database

A time series database is a specific database optimized for the storing of data indexed by time, such as temperatures, transactions, energy consumption, mobile phone activity, or financial market data, such as stock quotes or currency prices.

In the case of rapidly changing data, an RDBMS may not be practical as the number of rows may be immense and queries may be difficult for historical data. Apart from that, you may want to wire specific time functions (information condensation, such as summing or averaging over different time intervals and time zone conversions) close to the data. Often time series databases contain built-in time series arithmetic that works on a complete time series including multiplication, addition, rescaling, or otherwise combining various time series into a new time series. In financial services specific time series functions would include correlations (e.g., of a stock with an index), returns (of portfolios or funds), and currency conversions.

5.2.1.7 OLAP

OLAP stands for *online analytical processing* and is designed to answer multidimensional forecasts. In forecasts of dividends, for example, *multiple* forecasters create forecasts *at a certain point in time* and *over a certain horizon*. Then the information dimensions are the specific stock for which a dividend is forecast, the specific person doing the forecast, the moment in time when the forecast was made, and the period for which the forecast is made. To process information in more than a few dimensions, OLAP cubes are used.

An OLAP cube means a multidimensional array of data and the word *hypercube* is used when the number of dimensions of the data is more than 3. Information in the cube can be *sliced* and *diced* to get specific insights. Slicing means picking a subset of the cube by selecting a single value for one of its dimensions, for example, all the dividend forecasts for a specific stock or all the forecasts made by a specific individual. This creates a new OLAP cube where the number of dimensions is one less. Dicing is essentially a query on the data; it means making a subcube by picking specific values of different dimensions, for example, picking the forecasts made for all the stocks in a certain industry, made for, say next year only, made by a certain subset of analysts.

OLAP cubes and the slicing and dicing are often done for customer segmentation, for example, by zip code, credit score, age bracket, marital status, and disposable income bracket (Fig. 5.11).

OLAP is used for semireal-time data, for example, to come with recommendations and intraday reports. It is associated with digging into the data, roll-ups, and mining history. OLTP on the other hand is used for operational systems that handle transactions. This means feeding a set of instructions to process. Gartner coined the term HTAP (hybrid transactional/analytical processing) for the combination of the two sets of capabilities.

5.2.1.8 In-Memory DB

When databases primarily rely on main memory for data storage, we call them in-memory as opposed to traditional databases that use disk storage. Using main memory for storage will reduce the seek time and increase performance. You will still need to commit the data to a storage medium but that can take place later on.

5.2.2 NoSQL Databases

"Not Only" SQL databases are a new category of storage technologies that take a different approach. In NoSQL the mechanism for modeling and storing data moves away from the tabular form in RBDMS. Development of these technologies was spurred by the Web 2.0 companies. Primary considerations were 24/7 uptime, horizontal scaling, coping with very large numbers of users and volumes, and coming from retail platforms where short-term inconsistency was less critical than in financial services.

Some operations are faster in NoSQL because of the different underlying data structures, such as key–value, wide column, graph, or document. Compared to RDBMS, NoSQL databases can scale more naturally to large data sets. Also, they are more suitable to managing changing data because you do not need to define a relational model in advance.

The CAP theorem (or Brewer's theorem) states that it is impossible for a distributed computer system to simultaneously provide the following guarantees:

- consistency (all nodes see the same data at the same time);
- availability (every request receives a response about whether it succeeded or failed);
- partition tolerance (the system continues to operate despite arbitrary partitioning due to network failures).

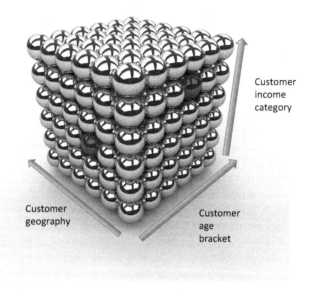

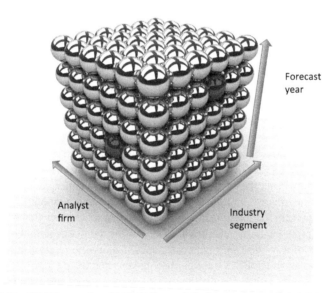

FIGURE 5.11 OLAP examples. *OLAP*, Online analytical processing.

In other words, you have to choose your storage model based on which of these three elements you are willing to sacrifice. As stated previously, in distributed computing the BASE properties replace the ACID properties of relational databases.

In NoSQL compared to in RDBMS, consistency is sacrificed for speed and availability. Instead of immediate consistency there is the notion of eventual consistency in which database changes are propagated to nodes eventually: that is, there may be a short period of time in which stale data or inaccurate data is returned and there may be some lost writes or data loss if individual nodes go down.

Organizations adopt NoSQL databases because of the need for speed, scaling options, continuous availability, and cost-effective solutions and also to be able to manage new types of data harder to manage in RDBMS, such as unstructured data.

NoSQL databases are often categorized into four families:

- **Column store.** Each row stores a flexible number of columns. Data is partitioned by row keys.
- **Document database**, also called semistructured data. The key uniquely identifies the document that can be encoded in XML or JSON. A document in a document database is similar to a record in a relational database but the structure is far less rigid and docs not adhere to a standard schema. In RDBMS every record in the table has the same form, plus you need to define the data type for each field a priori. In a document database there may be no internal structure that corresponds to the concept of a table, and the fields and relations generally do not exist as predefined concepts. Documents can be organized into collections and groups. An example of a document database could be in datasheets for financial products (as in the Key Information Document in the Packaged Retail and Insurance-Based Products regulation in the European Union; see http://ec.europa.eu/finance/finservices-retail/investment_products/index_en.htm). You can model tables to represent product information or you can directly store factsheets, key investor documents, or prospectuses in XML format in the database.
- **Key–Value store.** Row-oriented data storage of simple (key, value) pairs. For every row you can store different properties, unlike the fixed tabular form of an RDBMS. Key–value stores treat the data as a more opaque collection that can contain different attributes for every record. In a relational database the structure of the table and the data types of the column are all predefined.
- **Graph database.** Storage and retrieval of data that is stored as nodes, properties, and links ("edges"). Nodes are objects, such as customers, products, or accounts. Properties are pieces of data related to notes, such as customer address, customer credit rating, product risk score, or account number. The edges represent the relationships between nodes or between nodes and properties. Graph databases are similar to key–value databases with the additional notion of a *relationship* between nodes. Again, the relationships are not predefined as in RDBMS through foreign keys and are no a priori restricted. In other words, similar as for the data itself, new types of relationships can be added on the fly. A graph database focuses on the information in the edges and is built to look for patterns in the connections and interconnections of nodes, properties, and edges (Fig. 5.12).

There are many use cases of NoSQL technologies. Given the origin in Web 2.0 companies, these focus on online applications, such as:

- online web retail (transactional platforms, online shopping carts);
- real-time data analytics;

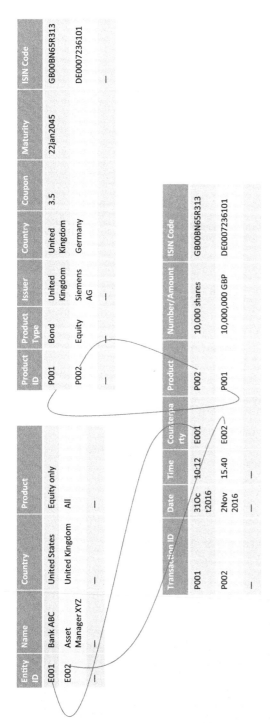

FIGURE 5.12 RDBMS representation and graph database representation of a bond transaction between 2 counterparties.

(continued)

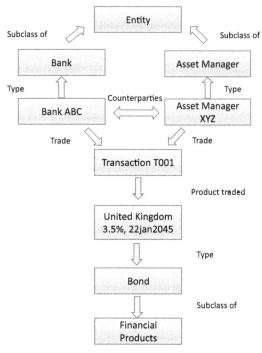

FIGURE 5.12 (cont.)

- social media capture and analysis;
- web clickstream analysis;
- time series feeds (financial or other time-based data);
- device/sensor/data "exhaust" systems in the internet of things;
- distributed transactional applications;
- write-intensive transactional systems.

Apart from these there are increasingly offline use cases that include:

- buyer behavior analytics;
- customer recommendation and upsell or cross-sell output;
- fraud detection and antimoney laundering ("AML");
- risk analysis and credit scoring;
- supply chain analytics.

5.2.3 Data Curation

The process of collecting, integrating, and verifying data around the storage is called *data curation*. Typically, you would want to separate the curation (loading, reformatting, validation) of data from the online store where data is accessible to end users. The reason for that is that you want to control what you make available (only approved data) and that you also do not want ad hoc queries to interfere with a scheduled process of loading, reformatting, and cleansing to arrive at updated master data in time. Also, at the place where the data has been processed prior to publication you want to keep audit information on who changed what and when and you want to store the *lineage* of the master data: that is, which sources contributed what.

Enterprise data management (EDM) is the practice that overlays these storage technologies with a data sourcing, quality management, and controlled distribution process. EDM tools refer to a category of systems that focus on the preparation of master data that then typically ends up in a data warehouse or in marts for further distribution. Common features of EDM solutions include:

- integration with internal and external data sources;
- mapping rules to combine different sets of content and to recast them into a common data model;
- validation rules to screen data sources and to detect anomalies, missing values, and inconsistencies;
- distributing data sets to downstream applications depending on subscription lists;
- keeping an audit trail on data changes and providing lineage to trace the origins of data values;
- monitoring incoming and outgoing data feeds against operational SLAs;
- providing a common data dictionary and workflow processes for data cleansing (given the need for joined-up data, there is increased attention for a common data dictionary and lexicography; at this point these tools overlap with data governance tools discussed earlier).

5.2.4   Data Warehouses Versus Data Marts

Data warehouses and data marts do not refer so much to specific storage technologies but rather to the scale of use of a database. The principal differences are shown in Table 5.2.

Often a data warehouse is the central large-scale data repository that feeds subsets or slices of its data to data marts. Data marts will be oriented to a specific business line or team. Data in the data warehouse will rarely be accessed or

TABLE 5.2 Data Mart Versus Data Warehouse

Data mart	Data warehouse
Focuses on department usage	Enterprise-wide usage
Typically one subset of information and use case, such as customer reports or financial reports	Contains information on different subject areas
Sometimes summarized data only	More detailed data
Data modeled in simpler model, such as the star schema	Richer data model
Integrates data from sources focused on specific functional domain	Large number of disparate data sources

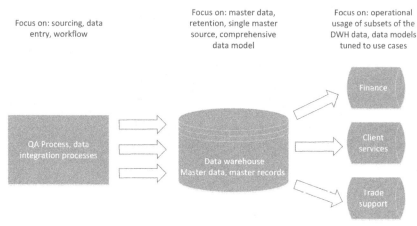

Focus on: sourcing, data entry, workflow

Focus on: master data, retention, single master source, comprehensive data model

Focus on: operational usage of subsets of the DWH data, data models tuned to use cases

Three steps:
1. Data sourcing, quality management
2. Data retention, master record for enterprise usage
3. Reporting databases, data marts for departmental usage

FIGURE 5.13   Data sourcing, data warehouses, and data marts.

altered directly. Data marts on the other hand will be owned by specific departments that could alter the data in their data mart and can sit downstream from data warehouses (Fig. 5.13).

A risk in large data warehouses is that it can be hard for the business to recognize their data. If a data warehouse becomes a wall between data and users it will lead to loss of data ownership.

5.2.5   Data Lakes

One of the main data management and therefore IT infrastructure challenges is to join up all the data in the organization. Data marts institutionalize silo approaches and hardwire today's question in infrastructure. In reality you do not know what questions or queries will come tomorrow so you need the ability to connect different data sets cross-department, cross-product line, cross-risk category, and cross-jurisdiction. Any boundary imposed by a database will have to be crossed. Data marts are the most efficient means to solve BAU reporting for a short moment in time, but no more than that as you can ever respond to only predetermined questions with marts. Also, data marts and data warehouses process data before committing it in storage—so access to the raw data is lost forever. With interpretation and molding of data comes loss of information and it could be that interpretations of data change over time.

Simply put, a *data lake* is a large-scale storage and processing engine that provides storage for any kind of data, enormous processing power, and the ability to handle many concurrent tasks. A data lake is essentially the data warehouse with BI tooling in the era of big data.

Whereas data marts cater to use cases or departments and data warehouses refer to a predefined scope of data, the data lake is a single store of all data in the enterprise. This can include raw data from source systems but also processed data used for reporting, analytics, and visualization. The data lake can include data from all types, including structured data from relational databases (represented in rows and columns), semistructured data (in markup, such as XML and JOSN but also comma- and tab-separated files), text data, such as emails and documents, and binary data, such as images, audio, and video (Fig. 5.14).

The data lake sounds like a nirvana: all of the data with lots of processing to extract value out of it and uncover patterns. But what does it mean in practice? Does it not just simply mean that you postpone doing anything to the data

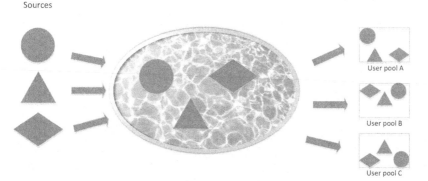

FIGURE 5.14   Data lake versus data warehouse. *ETL*, Extract, transform, and load.

TABLE 5.3 Differences Between Data Warehouses/Marts and Data Lakes

DWH/Mart model	Data lake
RDBMS	Different storage models
Light queries	Heavy queries
Data precurated for anticipated usage in departments	Cross-silo discovery
Data cast in tables and data types	Unstructured data

until the point when you need it and then you have to start from scratch building the analytical capabilities? As Sean Martin, CTO of Cambridge Semantics, said, "We see customers creating big data graveyards, dumping everything into HDFS (Hadoop Distributed File System) and hoping to do something with it down the road. But then they just lose track of what's there" (quoted in http://www.pwc.com/us/en/technology-forecast/2014/cloud-computing/features/data-lakes.html; also see http://hadoop.apache.org/index.html). Putting all the data together and postponing the predefined molding and interpretation of an RDBMS will preserve flexibility but will shift the discovery and query effort to the data discovery intelligence that you put in the queries (Table 5.3).

5.2.6 Social Media Meets Banks

One of the paradoxes about unstructured data and financial services is that financial services *is all about structure*. For the first applications of big data and NoSQL, it is probably best to look at areas closest to the online retail space where they originated. These application areas include mobile banking, transaction analysis, and in general recording all financial behavior of a customer in a time line, similar to the key data organization in Facebook and LinkedIn. [HSBC: http://www.hsbc.com/about-hsbc/hsbc-in-numbers (47 million customers). Citibank, http://www.citigroup.com/citi/about/citi_at_a_glance.html, estimates around 200 million customers. ICBC states 465 million personal customers, http://www.icbc.com.cn/ICBC/About%20Us/Brief%20Introduction/. By comparison, LinkedIn had 433 million users in Q1 2016 and Facebook 1.65 billion monthly active users in Q1 2016.]

A big difference with banks is that for social media networks the world is one big domestic market. Although local privacy laws can slow down social media firms here and there, the local regulatory differences and the cost impact of operating in different jurisdictions between these retail information collection and advertising platforms and banks are huge.

5.3 BIG DATA TECHNOLOGY FOR FINANCIAL INSTITUTIONS

One of the main developments in data management has been the increased scope of data that is processed and the volume of data. Normally, we see only the bits of data we directly interact with: the data in reports in front of us, the data we look up in a business application or on the internet, and whatever is put to our attention via email and social media.

Most of the data stored and processed, and most of the data that is underpinning the end results we see in documents, emails, or websites, is below the surface. This big iceberg under the sea level can pose risks. How do we expose it? How do we lower the sea level so to speak, to expose the rest of the data? And how do we avoid the refined, processed, verified data from drowning in noise? How do we shed light on the "dark data" prevalent in our enterprises, processes, and business systems? Because of the massive growth [IDC estimates that the data volume will grow between 40% and 50% per year; by 2020 the volume will be 40 Zettabytes (ZB); source: www.idc.com] in captured unstructured or semistructured data, database and data analytics capabilities have rapidly developed. At the same time, technologies like deduplication, compression, and analysis tools are needed to prevent seeing the wood for the trees.

Big data technologies have started to shed light on *unstructured data*, data that previously yielded its value only through large-scale human effort. Thanks to new computing capabilities, the processing of unstructured data is automated. A lot of change in storage technologies has been pushed by the sheer availability of data. The main properties of big data are often captured by the "3V's" (the 3V's concept was introduced by Gartner analyst Doug Laney in a 2001 MetaGroup research publication, *3D Data Management: Controlling Data Volume, Variety and Velocity*), later extended with several additional "V's":

- *Volume.* This is simply the amount of generated and stored data. The size of the data determines the value and potential insight. Often we have moved from selections of data to using every data point available in analysis. In storage we have almost stealthily moved up many orders of magnitude from megabytes and gigabytes to terabytes and petabytes. Similarly, the number of transactions handled by a transaction processing application has also gone up orders of magnitude in the past decades.
- *Variety.* The type and nature of the data. The range of data sources and types has gone up and this helps increase the insights that can be drawn. From data premodeled in database tables we have moved to processing unstructured text data (news, emails) and binary data (images, sound, and video).
- *Velocity.* This refers to the speed at which the data is generated and processed to meet business demands and challenges. Processes increasingly move from batch to multiple time slots per day, to near-real time to real time. We have seen this in the financial industry with shortening settlement windows for securities transactions and payments.

The challenges of data management when data has these properties stems from the simultaneous growth in all three, rather than just the volume that would be a case of scaling existing technology. Additional V's have been added to the original model:

- *Variability.* Inconsistency of the data set can hamper processes to handle and manage it. The range of values typical of a large data set has gone up.

- **Veracity.** The quality of captured data can vary greatly, affecting accurate analysis. Analytics has to take profiling and quality assessment into account.
- **Value.** The need for valuation of enterprise data.

We discussed NoSQL databases earlier and why organizations adopt them. The growth in data volumes, the possibilities to glean value from data not fit to be represented in relational databases, and the need to service large client bases in near-real time are all part of the answer. Developments in databases and analytics make more information to be captured earlier in the supply chain and vastly more information is available to be processed and analyzed. We can sum up the main trends as follows:

- More data is captured, preserved, and analyzed in unstructured format. This includes chats, messages, emails, tweets, geospatial information, audio, and video.
- As the range of data collected increases and we move away from using only carefully curated data in relational databases, metadata that provides context on where the data was sourced and what the confidence is in the accuracy will become critical.
- Concurrently in financial services, more records *have* to be kept and disclosed when demanded for investor protection and transparency to help regulators with market surveillance functions.
- Due to increased competition and changing customer shopping behavior, banking and investment management become more like other retail services, apart from the protective walls of regulation.
- As more client interaction shifts to online portals, BI tooling is used to process this digital trail and to come to personalized services. Financial services firms need to make more use of these tools to gauge client behavior and influence buying decisions.
- The nature of controls and the way internal processes and data are checked will change. The process used to be to take random samples and inspect these. Now that data volume processing is no longer the constraint, the entire data set can be subjected to analysis to self-discover the patterns and anything that can be fit into a set of "standard" profiles. Each and every transaction or data interaction can be scanned and put in its context to look for exceptions. This also means that risk, audit, and finance processes can become much more continuous and granular.

The main use cases for big data technologies across all industries are enhancing customer experience and more targeted marketing. Specifically in financial services, improved risk management is a use case. The ROI is still unclear in many cases as the industry is moving at the moment from labbing and experimentation to embedding big data technologies in business processes. How to get value from these technologies is still the top big data challenge and a large part of firms do not yet know whether the ROI of big data investments will be positive (Gartner 2016 big data report, see http://www.gartner.com/newsroom/id/3466117).

5.3.1   Developments in Analytical Capabilities

The large amounts of data captured and kept in NoSQL databases are useless without the tools to retrieve and process the data: *search* technology and *analytics*.

To avoid drowning in the abundance of data, *semantic search* seeks to understand the *intent* of the person behind the search. For this the contextual meaning of terms is needed to give back relevant results. Tools become increasingly better at second-guessing the *intent* of a user by using a rich information context in the user profile—the log of every interaction with that user to date plus whatever external information on that user is available.

In *analytics*, there is a logical progression of the role and direct impact of analytics that specifies the extent to which processes are directly influenced by analytics and the extent to which the human factor changes. Analytics can be *descriptive* and summarize what has happened. It can also be *diagnostic* when it provides guidance to find errors or improve processes. In scripted client interaction, such as that in call centers or via online portals, *predictive* analytics is used to suggest appropriate responses, guide the discussion, and suggest promotions and offers to clients. Predictive analytics allows operations to scale from local branches to central fulfillment centers. It also reduces the dependency on trained staff. At the moment when analytics becomes *prescriptive*, it can directly act on the conclusions.

Analytics has moved to machine learning and into artificial intelligence. Because in financial services most assets are information assets and not physical assets, the potential for AI tools is large and includes:

- operational improvements due to opex reduction (automation of data cleansing, reduction of legacy);
- client service (lower customer churn, finding optimal asset allocation in robo advice);
- front and middle office: arbitrage opportunities and investment decisions;
- risk and compliance: derisking via automation, proactive monitoring, and surveillance.

At some moment in the evolution of AI, the role of humans and software can be reversed. At that point the analytics will no longer be the surveillance function on the actions of humans but humans will watch the actions of the algorithms in the programs and intervene when necessary to further develop the algorithms. KPIs will have to evolve to support this development. Further developments are likely for automated, repeatable processes to more directly talk to each other with human surveillance only.

The quality of collected raw data will vary but the widespread adoption of analytics will push up the data quality. Confidence measures on the data and its sources will weigh in on the use of the data and data profiling will help form an opinion of the quality. There will be an increasing focus on differentiating the signal from the noise. Structured data, such as customer records that has gone through a curation process and strict controls when it entered the organization will be used as an anchor, as a foundation for the unstructured data. You can use unstructured data for financial product development but you need to get the basics right to be able to take advantage of the mining of the unstructured data.

5.3.2 Use Cases in Financial Services for Big Data Technologies

The use cases in financial services of big data technologies are limitless and it is easy to get creative. The implications of increased data exploration and embedding of predictive or prescriptive analytics for the future of financial services in terms of client interaction and tailored product creation are massive for the staff numbers needed in financial services. Firms have many more means to continuously track the behavior of their clients, record and use all market activity, and automatically come to personalized advice in a fully digital and mobile world. Financial service providers will necessarily become more like Amazon when it comes to recommending products, collecting feedback, and generally using previous customer interaction to improve the relevance of the platform. Examples include:

Retail client time lines to become a digital financial life coach. Keep track of all activities pertaining to each client. This will include all client communications, all information volunteered by individuals on public social media, and all transactions whether with your firm or transactions you can infer from, for example, social media posts. Retail client activity tracked will include when they visited your site, what investments were made, bank payment and credit card history, and shifts in spend categories. From this life events including children, move, promotions, marriage, and so on can be inferred, in short everything that would traditionally be volunteered in a face to face meeting with a trusted account manager. All this information will lead to a digital financial life coach offering personalized financial solutions including savings solutions, investment plans, insurance, and retirement planning. Recent developments in this area include the emergence of *Robo Advisers*, online wealth management platforms with automated, algorithm-based portfolio management without human financial planners.

Market logs for research and investment strategies. Keep complete transaction and quote price histories for different trading venues side by side. Regress every time series with everything else to look for patterns and new ideas. Slice and dice the data in different ways, by industry, by country, and by financial instrument to come to new indices, new sets of financial products to craft personalized investment solutions.

Credit risk assessment for every client from capital markets customers, and the financing of small and medium enterprises to retail clients. A time line of all activity by counterparty will be kept with rollup through a legal hierarchy from individuals up to the parent company. All transactions but also all information including authorizations, emails, and other documents will be kept. Nontraditional sources, such as the social media presence of companies, activity in peer-to-peer payments, and sentiment analysis on their products will provide new credit indicators to come to a predictive score. For example, reviews on Uber and AirBnB provide different proxies for creditworthiness and trustworthiness that are tracked continuously. For SME, reviews of their products and indications of retail buying patterns can help come to a credit assessment. As more retail business but also B2B shifts to online channels, more data is collected immediately and continuously. People like to discuss what they buy on Facebook and Twitter making for instant data collection. You no longer need to go through an elaborate data collection exercise before you can make forecast on shipping volumes and user satisfaction and thereby be faster in revising creditworthiness assessments.

Data sources inventory for cost management and compliance. A record of which data per data source was used to prepare data sets at what cost for each destination plus quality metrics. The complete lineage to be able to reconstruct the picture around a decision at any point will be kept. There are regulatory reasons here but also simply a safety net to have data supplied by a customer at hand in case of a dispute about fees.

New technology and the dominance of Web 2.0 companies in commanding the attention of customers via mobile devices leads to challenges for traditional financial service providers. Banks have to ask themselves whether it is only the banking license (and the provision of credit) which makes them unique and whether there will be any regulatory-walled gardens left for them.

An important point to note is that most of the data on retail clients that can be used to personalize and improve financial services is volunteered. The first condition for that, and consequently for the ability of banks to turn into digital financial life coaches, is *trust*. Trust is the reserve currency and mobile technology with improved data management can help restore that in an era of damaged brands. Mobile platforms and a tailored online experience have largely replaced the branch network with individual account managers. Financial services is the business of data manufacturing and good, comprehensive, accurate, timely information communicated to customers in an attractive mobile platform creates trust. The good news is that customers do not mind self-service via apps and sites, as long as the experience is good and the data is trusted. This creates room for new intermediaries.

As business shifts to mobile platforms and we move from scripted interactions to algorithms and prescriptive analytics, the number of humans needed in financial services processes will go down but their impact on monitoring processes, putting data to work, and developing algorithms will increase. Knowledge of big data technologies and what they can do will become more important.

5.3.3 Conclusions

Big Data technologies cast a wider net and let more data weigh in on decisions. Big data is maturing and moving from overcoming the challenges of data volume and variety toward getting value and getting anchored in business processes. Strategies are being built around different use cases, business goals, and outcomes, but primarily around improving customer experience and coming to personalized financial product offerings.

The maturity level of analytics is increasing through the integration of new relevant data sources and taking quality measures into account. Analytics is becoming closer to real time and will shift from being explanatory and diagnostic to playing a larger role in live business processes. It will become more forceful in influencing decisions up to the point of direct intervention. The role of the human will shift to monitoring these processes, intervening and improving the algorithms when needed. Consequently, the impact and data management firepower of any single individual will become larger. Staff will be much more productive and, in that sense, human capital will only increase with technological progress. The need, however, for staff to operate as responsible *data citizens* becomes all the more crucial.

5.4 DATA SECURITY

The first wave of digitalization was that of automation of internal processes. This stayed inside the organization, inside the walled garden of a private data center. The second wave started with the digitization and automation of supplier–client interactions. Following B2B e-commerce platforms, this digitization has spread to every retail client with internet access and a mobile device. This means every client has an online transaction platform 24/7 at their disposal and expects to be able to transact with their suppliers. This encounter between the formerly walled garden data center and the public internet leads to vulnerabilities to different kinds of cyberattacks. This problem is exacerbated by the trend of "bring your own device" (BYOD) in which personal smartphones and tables directly access the network of a firm.

One of the paradoxes here is that consumers are often very conservative toward banks when it comes to providing personal information yet freely volunteer private information to social media. This could be simply a matter of matching their behavior with the trust they put in the company they provide information to. The brands of many financial services firms are still damaged and the image of social media companies is far better. In a world where 99% of transactions and traffic goes online, brand is more important than ever. Brand damage and corresponding indirect financial damage is 100 times the direct loss from phishing and fraud.

5.4.1 Risks and Threats

Cyberattacks target computer networks, databases, and individual computes to steal, manipulate, or destroy information by hacking into them. They originate from individuals or organizations with the aim of stealing industry secrets, sensitive customer data, directly extracting money, or simply causing damage. The attacks range in scale from installing spyware on an individual PC to attempts to destroy complete corporate infrastructure. Methods employed include:

- Viruses. These are self-replicating programs that attach themselves to another program or file to reproduce.
- Worms. These are self-sustaining programs not needing other files or programs to copy themselves.

- Trojan horse. This is a program designed to perform legitimate tasks but also performs unwanted activity in a hidden way. Trojan horses hitch a ride with legitimate software.

The cost of cyberattacks is significant (cybercrime is estimated to cost 0.8% of world GDP; see Center for Strategic and International Studies and McAfee, 2014) and risks and threats from these cyberattacks include:

- loss of customer information, such as stolen client information used to access bank accounts and extract money (this can also include stolen account balance information used to extort clients or provide to tax authorities);
- loss of intellectual property;
- loss of pricing power through stealing financial information and product cost information;
- theft of personally sensitive information or diplomatic information (take WikiLeaks as a high-profile example);
- business disruption through hacking of websites and online platforms;
- legal and regulatory exposure if client data is stolen.

Potential future scenarios based on the level of cyberattack threats and the responses are summarized in Fig. 5.15 (based on Kaplan et al., 2015, p. 34).

5.4.2 Mitigating Actions

Responses to these threats range from general business process changes and IT practices to specific cybersecurity measures (this paragraph draws from Kaplan et al., 2015).

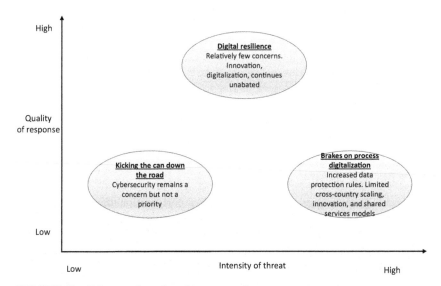

FIGURE 5.15 **Cybersecurity—three future scenarios.**

From a business process perspective, mitigating actions start with training your staff and making them aware of the risks. This can extend to enhanced background checks and monitoring as part of the HR processes. A firm will draw up more extensive security policies and will have additional demands on its suppliers, for instance, its software and IT services suppliers, especially the data centers. Specific security requirements include more stringent authentication and authorization requirements and automatic access right reviews built in.

Overall in IT, the security requirements on applications and on the network will go up. Digital right management in document management systems will be used to more tightly control access. USB ports have been locked down as have access to specific websites. The transfer of large files or files with unknown extensions is more closely tracked. Improved identity management is very important and will come via stricter password policies (strength of the password and frequency of change, additional control questions) and multifactor authentication (use of codes sent by text to previously supplied mobile number or email).

In authentication of customers accessing your portals there is experimentation with the use of tokens and biometrics. For instance, to avoid spoofing you can use the submission of a selfie and face recognition to authenticate someone on top of the use of tokens and passwords. Short videos and geospatial information can add further confidence in recognizing that the individual is who he or she says he or she is. Data is protected via encryption, both in stored form ("at rest") and especially when exchanged over a network ("in motion"). Data in databases is protected against loss by frequent backup. Conversely, sometimes personal data needs to be purged from all systems because of data privacy laws and consumer rights. This is not easy since data percolates everywhere from infrastructure to BI to EUDA systems and personal data may be stored dozens of times in a firm.

Mobile platforms have largely replaced the traditional branch network. Given the importance of trust in mobile platforms, and the increasing equation of the quality of the mobile interaction with the value of the brand, identification technology using biometrics and smartphone combination is a major area of innovation. The same technology combined with social media data can be used to directly challenge traditional financial services providers. Sufficiently trusted websites can take a "trusted third-party" function in peer-to-peer payments and crowdfunding as well.

5.5 BLOCKCHAIN

One very specific approach to data quality certification that promises increased security, efficiency, and lower costs is distributed ledger or blockchain technology. In this section we first provide definitions and discuss some of the current infrastructure challenges distributed ledger seeks to address. We then move on to list the plus points of blockchain and discuss different use cases focusing on master data. We end with challenges to the adoption of blockchain.

5.5.1   Definitions

Simply put, blockchain can be seen as a crowdsourcing of intelligence but with a consensus reaching process hardwired into it. *Distributed ledger* is a term that describes a database architecture where all nodes in a system collaborate to reach a consensus on the correct state of a shared data resource. *Blockchain* is a process of adding blocks of cryptographically signed data to form immutable records. Blockchains consist of time-ordered records, or ledgers, of transactions that are grouped into data structures called blocks. These blocks are secured from unauthorized modification using cryptography.

Each transaction processing node in a blockchain network contains a complete record of transactions and new blocks are added to the chain using a consensus algorithm that allows all nodes to trust that the block and the transactions within it have been verified as complete. All transactions can be viewed by any node on the network, making transactions transparent and auditable (Fig. 5.16).

Blockchain is best known for its support of the bitcoin currency and payment system and is gaining interest across financial markets. The bitcoin payment network arose as a peer-to-peer distributed platform to track and validate transactions without a central point. Key features of the blockchain approach include:

- The asset is built into the protocol.
- Transaction linkage—each transaction record is linked to the previous ones so that you can always retrace the complete history.

① A wants to pay B via an online network

② Transaction is packaged and digitally encrypted

③ Encrypted transaction is sent to all users of the online network

④ Transaction is checked via the network by all users and approved

⑤ Transaction becomes part of a collective shared ledger that is immutable by individual users thus protecting against fraud.

⑥ Transaction enters the blockchain. Payment completed. All users have a new updated copy of the ledger.

FIGURE 5.16   Blockchain steps.

- Every node adheres to a single standard as to how to store the ledger data and can have a full copy of the data.
- Blockchain uses decentralized consensus, which is a method to ensure that all transactions are validated and that validated transactions are added once and only once.

5.5.2 Current Infrastructure Challenges

Blockchain and distributed ledger technologies generally can modernize, streamline, and simplify the siloed design of the financial industry infrastructure. Distributed ledger technologies can play a role in overcoming limitations at different areas in the current infrastructure:

- Master data. Blockchain can reduce the multiple versions of the truth where every bank maintains the same data set over and over again through moving to shared data resources agreed upon by all industry participants.
- Vulnerability to technology threats. The current infrastructure was not architected for a world of complete digital interaction over the internet and is vulnerable to cyberattacks. Data can be quite easily compromised and because of the often redundant data maintenance there are many doors open if you want to compromise data.
- Complexity. IT systems and data sets have been gradually collated over time; in many larger firms we see technology from the 1980s to the 2010s coexist in one architecture. Poor integration between systems creates potential for confusion at the handover points.
- Not equipped for 24/7/365 processing. Most of the IT infrastructure dates from the end-of-day, batch-oriented world and mirrors a traditional day at the office rather than a globally connected infrastructure that must be up continuously to mirror the expectations of customers that expect to be able to transact via mobile platforms anytime.

5.5.3 Advantages of Blockchain

The main advantages of the technology include a potentially better, faster, and less costly way to process trades. This should lead to a reduced total cost of trade processing, reduced operational risk, and thereby lower costs for consumers. Important advantages include:

- A common shared version of the truth—every member in the network has a copy.
- All data is *encrypted* in the same manner and only the owner of the required keys can decrypt.
- Shared ledger establishes a network and a data standard.
- The transaction distribution model defines a method for active:active and continuous processing that is resilient to local database corruption.

5.5.4   Blockchain in Master Data

Blockchain offers a new approach to coming to a master data set or a single version of the truth. Rather than the in-house creation of master data that is repeated by each financial market participant, the process offers the possibility of coming to a common data set at industry level.

This takes place via mutual consensus verification protocols that allow a network to agree updates to the database *collectively*, with the certainty that the overall data set remains correct at all times without the need for a central governing authority. Master data could include security master data, legal entity master data, and information about trading venues, bank holidays, and so on. For instance, all the terms and conditions of a financial product that are currently taken out from the prospectus or bought from data providers could go into the public domain. Financial regulators, custodians, or central security depositories could become the stewards of this data.

Financial services is about accurate record keeping. Within financial services, Central Securities Depositories that keep records of securities ownership should be very interested in Blockchain. To quote a DTCC paper on blockchain technology: "DTCC's viewpoint is that basic industry master data is an ideal candidate for improvement using decentralized consensus, rule standardization and auditable change history. This information is used by the entire industry by definition, and the lack of consistency and quality is a recurrent industry problem. Further, this could be constructed in such a manner that multiple firms can be authorized as data submitters, there can be many data validators and the majority of users will be data consumers. However, it should be pointed out that some master data information that is specific to local laws written in nonprogrammatic legalese will be challenging to standardize in support of rules automation in the near future" (see DTCC, 2016; another source for this paragraph was Euroclear and Wyman, 2016).

5.5.5   Other Application Areas

Blockchain can not only be used in master data. Use cases can be found anywhere where an industry benefits from standards, trust, and scale economies and where there is risk of data corruption or ambiguity in processing. The paradox may be that any business is also about using and exploiting nuances (as in the reference to complex legal documents in the DTCC quote earlier) and that you need shades of gray to make money. However, in cost centers you want bright light and complete transparency to help keep your margins.

Potential effects and use cases of blockchain include:

- **Other data.** Apart from client or product master data, additional data, such as payment speed histories to gauge creditworthiness and trustworthiness of parties could be put in the blockchain. This could be similar to the rating and review system on Uber and AirBnB.

- **Simpler account structures.** Condense different levels of custody between bank, subcustodian, global custodian, and central securities depository. By flattening the account structure, redundancy and costs can be reduced.
- **Insight.** Regulators could get much more direct insight into very granular data if that is shared into a blockchain—something they now can get only via massive IT and data collection projects.
- **Clear ownership records.** Store all ownership data on securities and derivative contracts in a blockchain. Using digital signatures to sign transactions, unlock assets or cash, and then transfer ownership. Both sides can have pre-trade transparency that the counterparty can honor their side of the trade. Identity management associates a user identity with their holdings.
- **New issues.** Assets could be issued directly into the asset ledger. Securities could also be prestripped, that is, separate ownership rights to separate cash flows, such as the coupons and the principal of a bond. This could lead to disintermediation of parts of the banking industry. Also, if someone were to control a blockchain, it would be easier to enforce bans on short selling.
- **Smart and unambiguous contracts.** Distributed ledger technologies have programmable transaction capabilities and derivatives can be created as pre-programmed smart contracts to capture the obligations on both sides. Whereas in a bilateral contract you can think up anything, for derivative products to be used in blockchain there will be complete consensus on the terms and conditions in advance. A potential downside of smart contracts is that they can trigger liquidations if you build autoexecute rules into them. You will need to put in place circuit breakers to prevent ripple effects turning into crashes and to prevent an avalanche of smart logic kicking in the same direction (as in the 1987 stock market crash when portfolio insurance generated sell orders that accelerated downward momentum).

Adoption of blockchain could come in waves, from isolated proof of concepts to critical mass-required, industry-wide processes on the other end of the spectrum. Blockchain use cases could underpin some of the new shared services set up in the financial services industry to reduce costs. Examples include managing pools of collateral and KYC data services where the point is to arrive upon consensus documentation on new customers and other common data services. In the context of blockchain, KYC would mean the process by which parties are allowed into the community as a whole and allowed onto the distributed ledger as an actor and owner of financial products tracked via the ledger.

5.5.6 Challenges

Despite the intuitive appeal of blockchain technologies, there are significant hurdles for the widespread adoption in a market as heavily regulated as financial services. These challenges include:

- Different *regulatory requirements* on privacy are at odds with the distributed ledger technology. In some cases data needs to be physically maintained in a certain geography. This concept does not match with the idea of copies of the ledger being distributed to nodes on a global basis.

- Technologies, such as Blockchain that operate over the internet have no boundaries which raises the question of *governing law*. There is no single authority so it is not clear who owns the blockchain and how disputes will be settled. As the impact of manipulation will be higher, processes are needed to intervene in a blockchain, for instance, to change ownership after court rulings. Similarly, the flipside of there being no single point of failure is that accountability may not be clear.
- *Anonymity* is not always desired: regulators need to do market surveillance and maintain AML and antiterrorist financing controls. This comes down on the need to agree on a lead provider to hold responsibility (a common barrier for innovations). Regulators need to be able to intervene as per their legal responsibilities and will not accept technologies that prevent or materially complicate that intervention.
- It is not fully clear how you allow *trusted third parties* onto the ledger and how you integrate with legacy technology. If blockchain were adopted in core financial services processes, there will be many handover points that need to be crossed (referred to as *boundaries of trust*). These handover points need to be protected as these are typically the points at which data quality and integrity degrades. How many of these handover points will exist in any given organization?
- Some use cases yield benefits only when they operate at *sufficient scale* to get the network effects. This is simply a case of achieving critical mass and sufficient support before starting a blockchain project (Fig. 5.17).

5.5.7 Conclusions

Blockchain could lead to the fulfillment at last of an industry signed-off data model, something the industry never reached whether because of inertia or lack of will. An industry-shared understanding of financial products and counterparties in which every transaction is authenticated sounds like nirvana and would significantly reduce trade processing costs.

Indeed, blockchain could completely overhaul the notary function of CSDs and custodians in keeping track of who the beneficial owners of securities are. Also, it could overhaul the reference data world and the clearing and settlement processes, in fact all of posttrade operations. Regulators and tax authorities would no doubt demand immediate access to every transaction that would lead to a much more intensively, continuously policed capital market (Fig. 5.18).

5.6 CLOUD AND INFORMATION ACCESS

5.6.1 Cloud Models

Traditionally, data used to sit in the data center of the company. However, the number of options for storage have increased and include the offering of

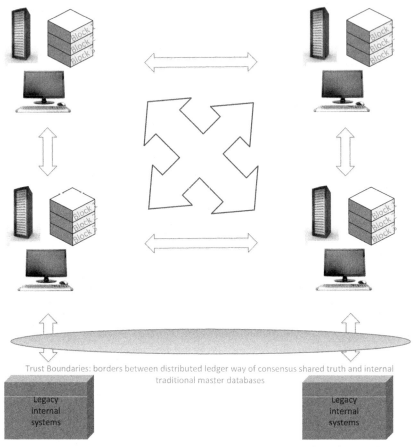

Trust Boundaries: borders between distributed ledger way of consensus shared truth and internal traditional master databases

FIGURE 5.17 **Boundaries of trust.**

software on a rental model ("Software as a Service") and the provision of public or private clouds with any potential combination possible too.

Characteristics of the most common models are listed in Table 5.4.

5.6.2 Data Center Requirements

The term "data center" applies to specially designed computer rooms. It can mean both a room inside your own building holding your servers or a complete warehouse with large arrays of servers that are rented out to different companies. Data centers come with standard racks to mount equipment, raised floors, and cable trays. With an increase in the uptake of cloud computing, business and government organizations scrutinize data centers to a higher degree in areas, such as security, availability, environmental impact, and adherence to standards.

Some countries have restrictions on where data can be stored, especially transactional and client master data, less so product master data. Unsurprisingly these tend to be countries with large private banking sectors, such as Switzerland

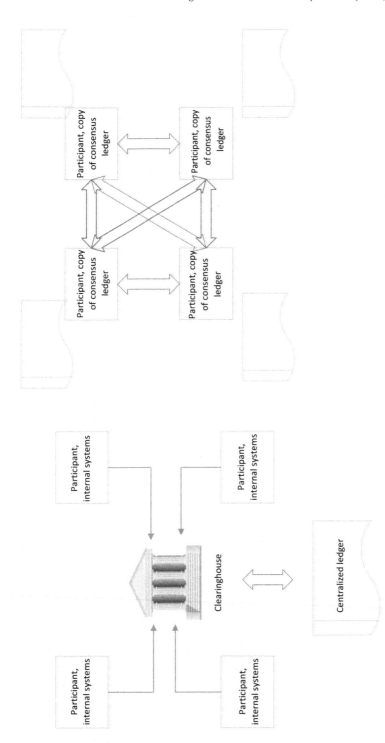

FIGURE 5.18 Distributed ledger as a clearinghouse.

TABLE 5.4 Different Cloud Models Compared

Model	Characteristics	Data management issues
Public SaaS	SaaS applications for specific business cases; Salesforce.com is the most famous example	Data access, US PATRIOT Act compliance
Public cloud	Public storage, suitable for low-level security applications without compliance issues	Compliance issues for confidential information. Could be used for master data and any nontransactional, noncustomer, and not commercially sensitive information
Private cloud	Infrastructure for high security concerns	Suitable for major part of infrastructure to handle all transactional business processes. High-end due diligence needed on supplier of private cloud given the criticality
Traditional hosting	If architecture rules it out, or mainframe-based applications running in a traditional data center	Often retail banking client processes and payment applications running on mainframes

and Singapore. Under the US PATRIOT Act, US officials can legally mandate any company that does business in the United States to reveal non-American data for investigative purposes. This is one of the factors in choosing data center locations (Table 5.5).

Data centers generally include redundant or backup power supplies, redundant data communications connections, environmental controls, such as air conditioning and fire suppression, and various security devices. Large data centers are industrial-scale operations using as much electricity as a small town. They are often located in colder climates to reduce cooling.

The "lights-out" data center, or dark data center, has almost eliminated the need for direct access by personnel. Because under normal conditions no staff needs to enter the room, they can be operated without lighting. Instead, all of the hardware is accessed via remote systems that monitor the operations. Apart from reducing staff costs and energy and being able to locate the center in remote locations, prevention of human interaction also reduces the chances of malicious attacks on the infrastructure.

5.6.3 Assurance Standards and Certification

Because of the rising use of external data storage and generally the involvement of more third parties in information management, independent certification of service providers is becoming more important. For business continuity and

TABLE 5.5 Data Privacy Laws

Switzerland

Federal Act on Data Protection, 1992 (DPA)
 	
Ordinance to the Federal Act on Data Protection, 1993 (DPO)
	• Provides an overall framework and deals with data protection using principles similar to those applied in other countries
	• Protects the privacy, interests, and fundamental rights of data subjects when their data is processed
	• Has central goals:
	  • The maintenance of good data file practice
	  • The facilitation of international data exchange by providing a comparable level of protection

United Kingdom

Data Protection Act, 1998 (DPA)	• Defines the law on the processing of data on identifiable living people • Lays down data protection principles to be followed and data subject rights • Enacted to bring UK law into line with the EU Data Protection Directive • Compliance is regulated and enforced by an independent authority: the Information Commissioner's Office

European Union

Regulation (European Union) 2016/679 on the protection of natural persons with regard to the processing of personal data and on the free movement of such data (repealing Directive 95/46/EC)	• Data Protection regulates the processing of personal data within the European Union • EU data protection reform started in January 2012 to make Europe fit for the digital age: new regulation adopted in 2016 and to be applied from May 2018

USA

Gramm–Leach–Bliley Act, 1999 (GLBA)
 	
Health Insurance Portability and Accountability Act, 1996 (HIPAA)
	• The United States made laws on "sectoral" basis:
	  • GLBA is applicable on financial services and protects personally identifiable information
	  • HIPAA is applicable on those organizations that process health information

US–EU Privacy Shield US–Switzerland Safe Harbor Framework	• To provide a streamlined means for US organizations to comply with the European Directive and Swiss Data Protection Law, the US Department of Commerce, in consultation with the European Commission (EC) and Federal Data Protection and Information Commissioner of Switzerland (FDPIC), developed a "Safe Harbor" framework and implemented a Safe Harbor *Self-Certification* program. The EU–US agreement was overturned following the *Schrems v Data Protection Commissioner* case by the European Court of Justice in 2015 leading to a new framework for transatlantic data flows. The US–Swiss agreement is in flux

operational risk management, firms need reassurance that the engine under the hood is sound and the organization is resilient enough to be there to service you and to have the operational means to honor the contracts. These certifications are often provided by accountancy firms that, apart from valuation of items on the balance sheet, also typically are there to control internal processes. Assurance standards include SAS70 and, more recently, ISAE 3402.

ISAE 3402 or "Assurance Reports on Controls at a Service Organization" (see http://www.ifac.org/system/files/downloads/b014-2010-iaasb-handbook-isae-3402.pdf) was published in June 2011 and is a standard procedure for documenting that a service organization has sufficiently adequate internal controls. ISAE stands for "International Standard for Assurance Engagements." Like earlier assurance standards such as SAS 70 and SSAE 16, ISAE 3402 defined Service Organization Control reports drawn up by service auditors that provide assurance to customers and service users of the certified organization. There are, two kinds of SOC reports:

- Type I: These SOC reports document a snapshot of the organization's controls at a specific point in time.
- Type II: These SOC reports document a period of time that is typically 6 months to show that controls have been managed over time.

The ISAE 3402 standard was developed by the International Auditing and Assurance Standards Board but it is also supported by other accounting standards boards. It supersedes the earlier SAS 70 and puts more emphasis on procedures for the ongoing monitoring and evaluation of controls.

5.7 IT MANAGEMENT AND BUY VERSUS BUILD CONSIDERATIONS

A larger financial services firm will have a substantial set of software applications to support its business processes. If you are lucky, these applications use the same databases and run on the same operating system but typically it will be a mix of MSQL, Oracle, and NoSQL databases plus Linux and Windows. Firms have to think carefully whether they favor buying applications from third parties or whether they develop in-house. If they do buy third-party products, they have to choose where on the adoption curve they want to be. This boils down to a choice between knowing with a high degree of probability you are making a suboptimal choice and taking a bet on something that can drastically change your capabilities (Table 5.6).

Build versus buy considerations can be found in Table 5.7. (For an overview of success factors in IT projects see the CHAOS report from The Standish Group on https://www.standishgroup.com/)

Models have been developed to describe world-class IT organizations or to rank the maturity level of software development organizations. Desired properties of IT organizations include:

- **Proactive.** Think ahead instead of staying in coping mode. This holds for day-to-day operations but also for the design and implementation of new

TABLE 5.6 Early or Late on the Technology Adoption Curve

	Cutting edge	Established products
Pros	New IP, cutting-edge functionality, and use of underlying new technology. Ability to steer the product road map	Proven technology Benefit from other firms' learning curve Services ecosystem available
Cons	High risk if small or no install base. Lack of services ecosystem, questions on support and longevity of the product	Risk of adopting out of date product No influence on product feature sets When will it be sunset—risk of product being retired

TABLE 5.7 Buy Versus Build Considerations

	Build	Buy
Pros	Get exactly what you need—there may not be a commercial product available that provides what you need Cheaper if your requirements are relatively vanilla No need to keep skills in-house Control change management process	Benefit from other clients' errors and input Third-party applications could use standards so you end up with more open applications If the install base of the vendor product is high, there are always skills in the market to implement and maintain
Cons	You may be reinventing the wheel and start on a learning curve a vendor's clients have already climbed Proprietary knowledge—hard to reengineer if people leave the company	Never an exact match to your business needs Suppliers can act unpredictably; they can be acquired, loose focus, or unexpectedly raise prices You may depend on supplier willingness to make enhancements to the software, or pay much more than you would have done if you controlled the code Be careful to prevent lock-in to specific entities, such as software vendors or their consultants

systems. Think ahead in terms of functionality but also in terms of expected future volumes, other nonfunctional requirements, and what the impact is on other parts of the IT infrastructure.

- **Service Oriented.** The IT department is a provider to everybody else in the firm. Features can turn into benefits when they can be used and when they are available to users. Keep the user in mind both in support and also in new application design.

- **Business alignment.** IT is a fundamental part of the operation and not just a cost center. The resources and skills need to be available in the department to deliver to the business.

Rather than aspiring to desirable properties of an IT organization as a whole, many financial services firms have recognized that a specific organizational model is needed to cope with the twin challenges of innovation and running a safe and efficient business as usual operation. To address this, some firms have adopted a *bimodal* IT organization. Properties of a bimodal IT model include:

- A secure and efficient business as usual operation, focused on stable production systems that run the operation with low operational risk. Change control procedures and operational risk control are strong here and the focus is above all on stability.
- A specific *sandbox* area focused on agility in which new applications from both established and start-up companies are tried out in short cycles allowing plans to 'fail fast' and making for a steep learning curve. Pilots can be run for specific countries, client segments, or financial products using these offerings. The focus is above all on innovation and new functionality. On success these products can be integrated into the BAU operation (Fig. 5.19).

The Capability Maturity Model (the Capability Maturity Model was developed at the Software Engineering Institute at Carnegie Mellon; see http://cmmiinstitute.com/) was developed to score the maturity level of software development. It has five levels of maturity and inspired a similar maturity model specifically for data management that was developed by the EDM Council. We will cover this in Chapter 6 (Fig. 5.20).

Agility:	Reliability:
• Focus on change • Business centric • Opportunistic • Fast feedback loops • Revenue, brand, customer profile, volume and experience driven • Short cycles • Agile development, fluid • Immediate and continuous	• Focus on governance • IT centric, business as usual • Long cycles, plan driven • Waterfall method • High performance • Existing client base • Conventional development • Think price/performance • Build to last

FIGURE 5.19 Bimodal IT model.

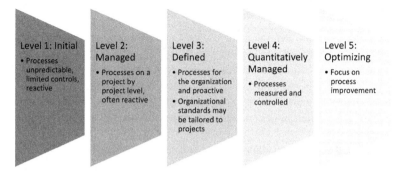

FIGURE 5.20 Maturity levels.

Gartner has a maturity model for IT infrastructure (see https://www.gartner.com/doc/527814/introducing-gartner-it-infrastructure-operations) that captures the main differences between a bad and a good IT department. Its levels are as follows:

- Level 0: *Survival*. There is hardly any focus on IT infrastructure and operations.
- Level 1: *Awareness*. The company realizes that infrastructure and operations are critical to the business. Some actions are taken to achieve operational control and visibility.
- Level 2: *Committed*. There is a move to a managed environment. Day-to-day IT support processes and project management aim to make the IT organization customer-centric.
- Level 3: *Proactive*. Efficiencies and service quality are achieved through increased standardization, and the development of clear policies and governance. Proactive, cross-departmental processes for change and release management are implemented.
- Level 4: *Service-Aligned*. IT is managed like a business in its own right with a focus on customers, to become a proven, competitive, and trusted IT service provider that could compare with external providers.
- Level 5: *Business Partnership*. The IT department acts as a trusted partner to the business. The business sees IT as increasing the value and competitiveness of business processes and the business as a whole.

Most IT departments would be around level 2 and some in level 3. Internal IT departments that are higher in this scale are rare in financial services. Partially this may have to do with the fact that most of IT would be outsourced: the infrastructure and IT operations would be managed by large IT services companies and most of the applications would be bought from commercial software providers.

Within IT departments, longer-term infrastructure and data management have tended to play second fiddle to application delivery. Shorter-term IT project mindsets can be at odds with achieving higher levels in maturity models that require a strategic, longer-term perspective and commitment.

5.8 CONCLUSIONS AND FUTURE OUTLOOK

What do the various technological toolsets and developments mean for data management in financial services?

A precondition for any information infrastructure is to have a clear agreement on data models and data dictionaries. If this is done well, it can be an enormous productivity gain. If done poorly, a lot of effort will be spent on reaching agreement *after* different data models and definitions have been hardwired into applications.

Increasingly, common data dictionaries are an industry problem. As we move from *front to back* integration, which is still focused on an *internal* view of transaction processing from deal capture to posttrade, alignment between the firms and its customers and service providers becomes critical. Digitization of business means electronic end-to-end processes and the integration effort does not end inside the firm. We are moving from front to back to *end-to-end* integration and continuous information exchange between principals in a transaction and any other part involved including brokers and other intermediaries as well as custodians, fund administrators, regulators, and CSDs.

Agreement on basic items, such as entity and product identification as well as clear naming conventions and domain definitions of product terms and conditions across the whole supply chain is critical for a smooth and cost-efficient process. This will also be the only way to measure and compare KPIs on processes end to end.

There is an increasing number of sourcing options for application development [commercial off-the-shelf software ("COTS"), in-house development, and the commissioning of third-party application development] and for IT operations. Key attention areas for sourcing departments are to get tools to preserve flexibility and not lock the data to the process. Preventing lock-in to specific formats or vendors and to always abstract from and control dependency on the specifics of any particular tool is also important. The sourcing department has to help find the optimal configuration of tools, operations, and services with as little constraints as possible that can act as the best *containers* for the business' secret sauce. In the bimodal model, sourcing departments will have to help spot promising pockets of innovation.

Social media creates rich data sets available for free or at low cost that can be tapped into using big data toolsets. Huge amounts of anecdotal information can be organized in databases and time lines on different axes including:

- Product view. What types of customers and how many are buying the product and what are they saying about it? How do they compare it to competing offerings? This can feed into real-time sales information and cost information and risk information. How is any external news impacting this?
- Client view. What do our customers do and think, where do they shop and go on holiday, and what are their commuting patterns? Impulsive expressions on Twitter take us one step closer to people's actual thoughts. What do clients talk about with whom? Are they changing jobs, getting married, having

children, or moving house? Enormous amounts of new information can feed into credit profiling.

The key challenge is to use the available information wisely, find actionable signals in the noise and address relevant use cases of integrating this in business processes such that customer interaction and experience is improved. Overstepping the use of personal information can backfire and in an age where the online portal makes up a large part of the brand can cause huge damage.

Big data and NoSQL database technologies that came out of internet firms are now finding application areas in banking and investment management. Cloud offers cost advantages as well as new sourcing options. Blockchain can both be used to guarantee consensus views on master data and ownership records and also change posttrade firms' business models.

In other words, technology will be not only be used to automate or achieve lower cost. Sometimes inadvertently, it is also a disruptive change agent for business processes. New firms with new business models and without the legacy infrastructure to carry will be built around it.

REFERENCES

Center for Strategic and International Studies, McAfee, 2014. Net Losses: Estimating the Global Cost of Cybercrime. Available from: <www.mcafee.com/us/resources/reports/rp-economic-impact-cybercrime2.pdf>.

Codd, E., 1970. A relational model of data for large shared data banks. Commun. ACM 13 (6), 377–387.

DTCC, 2016. Embracing disruption—tapping the potential of distributed ledgers to improve the post-trade landscape. White Paper. Available from <http://www.dtcc.com/news/2016/january/25/new-dtcc-white-paper-calls-for-leveraging-distributed-ledger-technology>.

Euroclear, Oliver Wyman, 2016. Blockchain in capital markets. Joint Report. Available from <https://www.euroclear.com/en/campaigns/blockchain-in-capital-markets.html>.

Kaplan, J.M., Bailey, T., O'Halloran, D., Marcus, A., Rezek, C., 2015. Beyond Cybersecurity: Protecting Your Digital Business. Wiley, Hoboken, NJ, p. 69ff.

Chapter 6

Data Management Processes and Quality Management

Chapter Outline

A Primer in Financial Data Management. http://dx.doi.org/10.1016/B978-0-12-809776-2.00006-5
Copyright © 2017 Elsevier Ltd. All rights reserved.

6.1 INTRODUCTION: METADATA CLASSIFICATION AND DATA MANAGEMENT PROCESSES

We have looked at data management from the perspective of financial data categories, business processes, and technologies and have discussed the challenges the financial services industry faces. The ability to tie information strands together and keep the overview can mean the difference between a well-oiled operation that allows for opportunities to *scale* in terms of new clients and products and an information jungle that has users drown in irrelevant information and endlessly wonder about the presence and trustworthiness of the information on which they depend.

This chapter focuses on aspects that can be used to define quality and fitness for purpose of information and on the metrics by which the efficiency of information sourcing, processing, dissemination, and generation processes can be assessed. We will discuss criteria with which to judge the quality of information and how to arrive at a well-managed information architecture.

We present examples and discuss specific benchmarks and Key Performance Indicators (KPIs). Processes can be hardwired into an organization and what often happens is that these processes are tweaked at the edges (spreadsheets, manual work) or abused (forcefully fit in new information into an existing data model) and therefore become increasingly less transparent.

We start by discussing data quality fundamentals and different aspects of information quality. In IT, the focus (and potential for organizational glory) has been on implementing new systems rather than on getting the most out of existing systems. We discuss business rules to measure quality aspects and different types of KPIs to track and act on quality indicators. The way most information architectures in financial institutions are set up means significant uncertainty or *entropy* in information and we will discuss ways to control this uncertainty, for instance, via data standards and data management maturity benchmarking. This is followed by an overview of impact analyses to gauge the ROI of investments in data management infrastructure and an overview of cost impacts of poor data quality. We discuss what the future holds for financial content and the way it is sourced and distributed.

As McGilvray (2008) puts it: "Information quality is the degree to which information and data can be a trusted source for any and/or all required uses. It is having the right set of correct information, at the right time, in the right place, for the right people to use to make decisions, to run the business, to serve customers, and to achieve company goals."

6.2 DATA QUALITY FUNDAMENTALS

Data quality management starts with an understanding of data decay and its causes. There are different reasons for data decay following the initial automation effort. These include:

- Changes in the data are not captured because they were unexpected, because new data value does not fit in the old system, or because the fact that the system has a separate independent data store is not appreciated.
- System upgrades. A system can be upgraded with data migrated to the new version. However, the new version may expect additional data items to be populated and may give them default values.
- New data uses. To keep the system in use, IT may abuse some fields in the database for special values that determine the behavior of the application.
- Loss of expertise. The people who designed the initial system may have moved on.
- Outsourcing/lack of training. "Your mess for less" is a phrase often heard when outsourcing is described. If someone new to a proprietary system does not receive adequate training, it is no surprise the quality of the information in that system will go down.
- Data migration. During data migration projects there is time pressure and quick fixes may be imposed to push the data into new storage models.

To slow down the decay of data it is important that it is easy to access and keep up to date. Second, a system should have users that know they are responsible for the data quality. Finally, the application can help with periodic prompts to confirm or update the value. This can be both applied to internal users and also used to ask clients to confirm their details.

IT developed toward automating end-to-end processes that cut across different firms' supply chains, instead of specific internal, departmental use cases. This means data integration is needed to overcome the potential for error at handover points and to reduce the opportunity for miscommunication and quality degradation. You need to know both the aspects you want to measure (data quality dimensions) and the most appropriate techniques to gauge the resulting business impact in order to prioritize data improvement efforts and make the business case for that.

Information is modeled, obtained, subsequently stored, and shared, maintained, used and eventually archived or sunset. The beauty of information is that it is not consumed but endlessly reusable. To assess the role of data quality in business processes, we need to understand its life cycle first. Quality is in the eye of the beholder.

A good data management process begins with an understanding of the constraints on the data in terms of security, content licensing, retention requirements, and privacy. This is followed by an understanding of the data itself: what is the data model, what are the relations between the elements of the model, and what behavior do we expect from the data? What are the permitted values and ranges of each of the data items, where and how do we want to test for validity, what is the structure of the data, and how are each of the elements defined in a data dictionary?

We require a data capture strategy and an assessment and feedback loop plan to keep track of data quality and to have change and improvement procedures.

Storage has become very cheap but storing the same information multiple times is a false economy. In data capture, first time right is a valuable principle and you need to be careful what you let in. To avoid cleansing downstream, focus on data quality and be critical on checking it as data enters the firm.

Once data is stored and maintained, governance procedures need to be in place. They define, for instance, the source for each piece of information and how information can be challenged, updated, deleted, or corrected. Governance specifies who can Create, Read, Update, and Delete information ("CRUD" rules) and the RACI matrix (RACI outlines who is Responsible, who is Accountable, who provides input to the decision in a Consulting role, and who is Informed but does not need to be consulted) for data changes.

When looking at an existing data set, we first need to understand the data. *Data profiling* is used to get to know the data and to get an initial feeling for potential quality issues. As McGilvray (2008, p. 118) puts it: "data profiling is the use of analytical techniques to discover the structure, content and quality of data." A profiling exercise includes looking for properties, such as:

- fill rate: how often are the fields populated;
- the number of unique values versus repeated value;
- frequency distribution across the values in a domain, for instance, if you have a 7-point rating scale and 99% of records fall into category 4 or 3, the scale may defeat the purpose;
- observance of naming conventions (for naming conventions, improving business efficiency and information on best practices to improve information quality see also the International Association for Information and Data Quality on http://www.iaidq.org);
- maximum/minimum values;
- the average age of the data;
- the average age of the most recent update per record;
- how often used, touched, and read;
- duplication of records;
- precision (of numbers) and variation in precision;
- use of masks for zip codes/postcodes, addresses, or other groups of information;
- quality of documentation of standards.

6.3 DATA QUALITY DIMENSIONS

Asking people how to define and measure quality can cause hazy looks and trigger philosophical and often heated discussions. In this section, we discuss different aspects of quality. Quality depends on the needs of the users at the receiving end; the dimensions on which you measure it will vary from department to department and from data type to data type. Quality has to be evaluated using the criteria that the data has to comply with. Metadata—data about data—may well be just as important as the data itself (Fig. 6.1).

FIGURE 6.1 **Different aspects of data quality.**

- *Speed.* The time dimension of data. When can you act on it? What is the decay factor in the value of it? In some cases (high-frequency trading) opportunities exists only for microseconds. Speed means getting information to the right place at the right time. How long does it take to service a client, set up a new account, or introduce new products driven by changes in the tax code?
- *Accuracy.* When do you need to be 100% sure of the accuracy, such as precise spelling or coding, for example, the full legal name underlying a Credit Default Swap or the certification of customer personal identity?
- *Consistency.* If you have different trading books for risk reporting purposes, you want to use the same exchange rates. Are you treating your various clients consistently or some more favorably than others? Firms can have many product lines that all serve the same customers and need to present one voice to the external world.
- *Authentication.* Has the source of the information confirmed the validity? Is a quote indicative or actionable? Has the trade been confirmed? Are there liability ramifications if the price or settlement instruction provided proves to be wrong? Is there a "twilighting" process in place to reaffirm validity periodically?

- *Transparency.* To what extent is the complete data lineage clear? For example, *what* information was used *when* around the decision making to take on a new client? Regulation increasingly demands *trade reconstruction capabilities*, meaning you need to be able to replay the decision-making process showing *who* knew *what* and *when* they knew it.
- *Synchronicity and order.* Are you preserving information for regulatory reporting in the right order, even if it happens at the microsecond level? Also, if you have to piece together a report and rely on cut-and-paste information and queries from various repositories, chances are that your information is not obtained at the same time and may be out of synch immediately.
- *Completeness.* Ensure that all values required by customers contain values. Pay particular attention to attributes considered "high risk" or which are used for key processes, such as identification or cross-referencing. Ensure that values that can be expected to change do not become stale. Is the full picture available? For example, you can have 25 covenants in a loan or a bond; you could have 100% accurate but nonetheless incomplete information if you report only 24. Another example on completeness can come from legal entity information where you will want complete information on the legal structure, and the guarantees and liabilities of an entity and any of its subsidiaries. Different use cases will require different levels of detail and completeness of the record. It is much easier to start the entity data provision for the onboarding process of a customer that may require 10–15 fields. The compliance function will need additional information on audit and documentation. Different levels of completeness of information could be:
  - *Research ready*, a subset of information is present.
  - *Compliance ready*, a potentially different subset of information on the entity is present.
  - *Trading ready*, the complete set of attributes is available and has been checked.
- *Relevance.* Is information filtered in useful ways? You do not want to be cluttered with useless information and want the useful bits extracted out of the torrent. Ways to address this could be to filter out the relevant updates only, to alert users only on a real update, and to prioritize or filter information offered by securities in which there is an open position, by the top 25 exposures, by credit rating, by price volatility, or by complex product exposure.
- *Control and maintenance.* Spreadsheets and macros represent the democratization of IT but can lead to dramatic control and maintenance issues. Macros and spreadsheets have put strong IT tools in the hands of the masses but without a management framework for version control, sharing, and reusing. Too much power can be concentrated in the hands of whoever understands the 180-MB spreadsheet.
- *Accessibility.* Does data reside on local desktops? Can everybody who needs it access it? Are access permissions, Chinese walls, and separation of du-

ties in line with content licensing terms and conflict of interest rules (e.g., between research and advisory functions or between corporate finance and sales and trading)? Data protection and privacy laws would especially be areas of attention when outsourcing or offshoring the collection or processing of customer information. Is the data (or access to it) leaving the building, leaving the country, or leaving the organization? Each of these events may or may not be allowed.

Another aspect of accessibility is how easily the information lends itself to automated retrieval and processing. Accessible to the human eye means something different than accessible via an API. Information may be *somewhere* in the institution, but there are often no yellow pages in place directing you to the relevant department. Which people have *backdoors* allowing them direct and privileged access to data stores, for example, database administrators, IT staff, and management? If the number of people with backdoor access is fairly large, the controls put in place for other users will become meaningless.

- *Usage restrictions.* Is the organization abiding by the content licensing terms in the contracts?
- *Security.* Is unauthorized access prevented? Is information encrypted? Are the standards of the institution with regard to security level of information (from public to confidential and shades in between) adhered to?
- *Service and support* around data. If you want to change something, what are the flexibility and the turnaround time? If you have an issue with it (query, error, or enhancement), what is the turnaround time in addressing it from your vendor, and from your internal data management team? Another aspect of data services is the availability; can you always access the data (this could be via *uptime targets* such as 4 or 5 "nines" that refers to 99.99% and 99.999% uptime, respectively)?

An obvious question is *how defendable is the data?* What we mean by that is the following: is it internal audit proof or, more importantly, regulation proof? Did we verify the data quality requirements in the applicable regulations? Solvency II has certain data quality requirements for insurance companies; the *Fundamental Review of the Trading Book* for banks has other quality requirements.

Linked to this is the question of data certification. Who has attested to the quality of the data? Can the providers of the data be audited? What responsibilities do commercial providers of data take in case of deficiencies? We can also summarize this by the term *transactability*, which is "a measure of the degree to which data will produce the desired business transaction or outcome" (McGilvray, 2008, p. 33). Transactability is, in other words, an indication of immediate usefulness with no further processing required.

What these different quality aspects really say is that information can be fit for purpose in different ways. Whether a user relies on that data for commercial activities, cost controls, or compliance can lead to different perspectives on quality. Furthermore, various subsets of these quality criteria are more relevant depending on the type of data (pricing or static, equity or commodity, counterparty or corporate action). For example, information quality aspects associated with the quality of order execution would include measures on effective spread, rate of price improvement or decrease (vs. a benchmark), fill rate of the order, and speed in turning it around.

There is no single right answer to the quality question. When it comes to defining standard data services, different users will want to embed different combinations of the factors mentioned earlier in their service level agreements (SLAs).

6.4 DATA QUALITY BUSINESS RULES

Business rules are criteria for making decisions on the data. They can flag erroneous or missing information, point out logical inconsistencies, and observe trends in data quality. Quality rules exist as a sieve to separate the exceptions from the BAU or the expected. Every human working with data will have in his or her head heuristics on what constitutes normal behavior of the data. The challenge is to capture all this knowledge and have it operate on the data efficiently so that human time is spent on the exceptions to the exceptions. In this section we provide various perspectives on business rules. We discuss transformation rules on data, rules by financial instrument type, an example of a set of staged quality rules that look at information quality at different levels of granularity, and example market data rules. First, let's look at common types of business rules for transactional or master data (see also McGilvray, 2008, p. 50).

Master data includes the following:

- Mandatory info: every bond must have a field coupon rate populated.
- Consistency: do we have date of birth of all our clients?
- Plausibility: if large cap stock moves more than double the percentage of its index, check the value.
- Proxy: if a price of a bond is missing, apply the return of a similar bond to yesterday's price of that bond.
- Permitted value: is the currency code in the ISO domain?

Transactional data includes the following:

- Restrictions: customers of a certain category cannot trade certain products.
- Preconditions: before allowing this transaction, check preconditions such as customer eligibility and available funds.
- Computation: calculate the net asset value of a fund.
- Inference: if customer does no longer get any cash inflow into his or her account, he or she may have lost his or her job or switched banks.
- Timing: transactions need to be reported to the regulator within a certain timespan after conclusion.
- Trigger: if customer trades approach his or her limit, send a text message to warn him or her about this.

Best practices in data quality management and data cleansing rules include:

- Proactively capture business logic in rule sets to screen data products: data should be assessed on its own merits, not based on whether multiple sources agree.
- Go to the original source where possible to capture business logic as well as data. By basing rule sets on empirical information about data items, many attributes can be completed even if the source does not provide them. This includes depositories and exchanges. Any handover point in a data supply chain adds a delay at best and quite commonly also degrades the quality.
- Go as far upstream as possible in the data management process to find issues. This starts by monitoring the arrival of incoming vendor data versus expected times and the size of delta files to see whether we get the expected amount of change.
- Look at data usage statistics including the freshness/decay of data, billing/cost allocation of data (whoever pays will shout first about the quality), and any license/vendor reporting requirements there are.
- Categorize data into different groups. Recognizing the interdependencies between security, legal entity, corporate action, pricing, and ratings data can identify issues that each set cannot reveal independently. For example, some static data should change only following a corporate action. If we detect a change to static data between two versions of the data item, we use the corporate action data set to check if the change is valid prior to raising an exception.

6.4.1   Transforming Information

Another category of business rules is *transformations*. Business rules often occur at handover points when information is extracted out of an application into a report—or when it needs to move from one application to another. Examples of specific business rules on financial instrument prices include:

- *Scaling* prices to express them in a common base, for example, a move from penny-based prices to British pounds, to express prices in the same reporting currency or to restate commodity prices into a common unit of measurement for financial reporting (example conversion in the case of natural gas can be to change the price from US$ per MMBTU into CAD$ per gigajoule).
- *Filtering* information through criteria such as AND/OR conditions, being in or not in a certain subset, and equal or not equal.
- *Arithmetical operations* such as adding, subtracting, multiplication, or division of information by a constant, or by a dynamic element (a currency rate). These are the basic operations such as creating a spread curve.
- *Implied prices from related instruments.* Economically identical instruments should have identical prices. Through insight into price determinants and risk factors we know that often pricing information is implied, for example, through related products that should move in tandem.

- *Implied additional pricing fields.* Often different price expressions exist for the same instrument and conversions take place between the two, for example, changes from price to yield for bonds or changes from price to volatility in the case of options. If different pricing measures are used for different instruments, these conversions make them comparable. Clean reference data (for options strike, expiry date, option type, risk-free rate, underlying, for bonds coupon rate, maturity date, payment frequency, day count convention, redemption price, and possibly optionality elements including call schedule and conversion price) is a requirement to be able to do these calculations accurately.

6.4.2 Financial Instrument-Type Specific Rules

Business rules serve to check the validity of the data against the quality aspects mentioned. They will depend on the type of data. An example of rules for different instrument types is listed in Table 6.1.

6.4.3 Staged Quality Process Rules

Business rules and quality management can be broken up in stages. For instance, a quality verification process can take place first at the source, then by looking at the instrument level, and then by looking at the field consistency level. Finally, data can be enriched to enhance externally sourced data (Fig. 6.2).

Example rules in such a staged process are as follows:

Source-level rules—Rules applied on the data vendor/exchange level. Examples of vendor-level rules are as follows:

- Identifier checks are present—valid length, checksums, format (ISIN, CUSIP, SEDOL).
- All mandatory fields are populated.
- The format of the data file is as expected.

Product-Level Rules—Rules applied for each financial product type. Some examples of product-level rules are as follows:

- Futures require a month, year code, and last trade date.
- Options require a strike price, put/call indicator, and a month and year code or expiration date.
- MBS Securities should have an issue date on the first of the month.
- Zero Coupon securities should not have payment dates, payment frequency, or coupon rates.

Field-type rules—Rules that are applied between fields. Examples of Field-type rules are as follows:

- For derivatives, the settlement date should not occur before last trade date.
- For listed derivatives, the contract size, tick size, and tick value should never be null.

TABLE 6.1 Example Data Quality Rules

	Equity	Fixed income	Exchange traded derivatives	Rates
Identification Information	Syntax checks on ISIN, CUSIP, SEDOL Cross-reference with vendor IDs Changes triggered by corporate actions (M&A) Link multiple listings	Syntax checks on ISIN, CUSIP, SEDOL Cross-reference with vendor IDs Changes triggered by corporate actions (M&A)	Exchange symbols, cross-reference to updated vendor symbols, housekeeping of universe	ID construction on attributes, cross-reference and check between IDs of different data providers
Income characteristics	Dividend info (expected corporate actions) Rights, tenders (unexpected corporate actions)	Expected changes (rate resets) and unexpected changes (put/call options), default	Contract specifications, link to underlying products	NA
Issuer data	Events on change of registered address, M&A, issuer ratings, link to entity/ultimate owner	Events on change of registered address, M&A, issuer ratings, link to legal entity data sets, ultimate owner	Exchange identification	NA
Pricing data	Quote consistency rules Plausibility rules against history, indices, other listings	Plausibility rules versus history, discounted cash flow "fair price," link with curve creation libraries, comparison with evaluated prices	Data consistency original sources, including other time series such as open interest. Corporate actions on underlying can cause jumps (depends on exchange policies)	Infer other currency pairs, implied deposit/forward rates. Return checks, plausibility checks
Internal data	Client-specific IDs, client-specific hierarchies, client processing info, tolerance parameters	Client-specific IDs, client processing info, tolerance parameters	Client processing information, local cross-reference, pricing policies, tolerance parameters	Preferences on quoting banks/brokers per product, currency, tolerance parameters
Tax and regulatory information	Withholding and transaction tax rates Instrument/entity taxonomies for regulation	Withholding and transaction tax rates Instrument/entity taxonomies for regulation	Withholding and transaction tax rates Instrument/entity taxonomies for regulation	

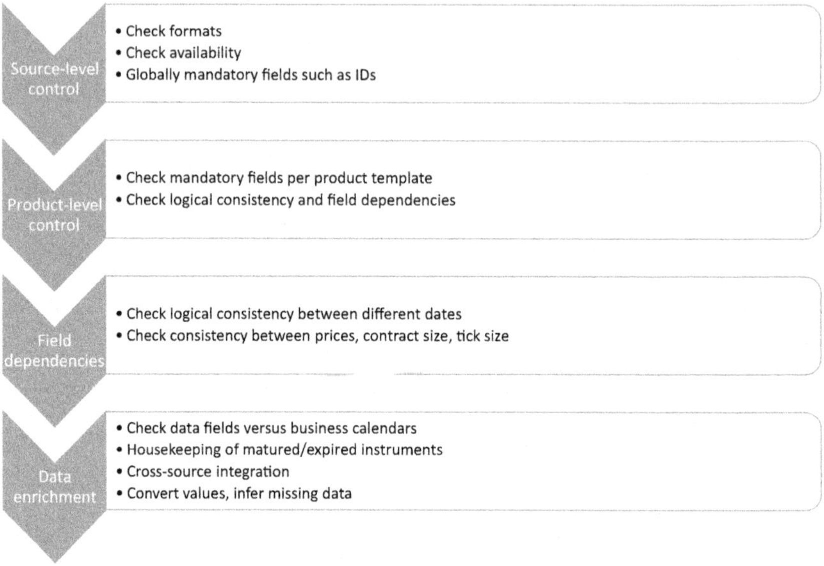

FIGURE 6.2 Staged quality management for financial instrument data.

- For fixed income, the first pay date should not occur before the issue date.
- For fixed income, if the payment frequency is not null or zero, then payment dates should be available.

Data Enrichment—At this level of the process rules are applied to enhance the current data files coming from vendors, exchanges, or clients. Example checks are:

- using the holiday calendar to make sure that dates do not land on weekends or holidays;
- manually calculating missing dates based on contract specifications;
- deactivation of matured/expired contracts;
- cross-vendor comparison of data;
- infer or convert values, for instance, for options translate between delta points and strike level to express moneyness.

6.4.4 Example: Market Data Rules

If we look at rules to validate for market data, a common data problem is that of filling in a missing value. This can be done in various ways using context knowledge:

- propagating the previous value;
- interpolating through time between the previous and the subsequent value;
- using statistical estimation techniques;

- using a proxy time series, either directly or by scaling the reference time series. For example, when missing a value for an equity, take the previous value of the equity and apply the return of the index times the beta to it. Alternatively, it could be proxied more specifically by taking the return of a peer group and applying that to the previous value. (Proxies are not confined to time series. They are also used in the area of credit ratings. In case an issuer is not rated, the issuer rating could be proxied by taking the rating of the senior debt.)

Other cases where a missing value has to be filled in can be found in completing a yield curve. Often, fresh data is available for part of the curve but data for intermediate points or the long end of the curve is missing. Tenor points can then be filled in using interpolation or extrapolation techniques. The specific interpolation method needs to be chosen with knowledge of the product underneath. Also, look to see whether the results make intuitive sense: when a spread curve or a credit spread becomes negative, something may be wrong with the benchmark. Additional common market data issues include:

- Stale prices. Quotes have been given by market makers that may be outdated.
- Low liquidity. There will not be many transactions and large open interest for many contracts, especially those far out or far in the money. Most of the value and trading activity will be around the at-the-money contracts.
- Large bid/offer spreads. In that case it makes a big difference whether the bid or the offer side is taken. Also, in some cases a dealer may quote only for one side and the *Bid* or the *Ask* quote will not be available.

The following validation functions can be run to find errors in price data:

- *Source comparison.* Compare two or more carriers of what should be the same information.
- *Semantic validations.* In some content areas, categories of filters with domain knowledge built in can be used: for example, for a retail CRM system logic that knows the format of addresses, that has knowledge of postal codes format and street name writing conventions in different countries. This includes spelling and capitalization rules in languages and common abbreviations of company legal forms (such as Ltd, SA, Inc, LLC).
- *Reference data consistency* and presence of information. Check various fields in the terms and conditions for consistency. If a bond is callable, a call schedule must exist and vice versa. In the case of an option, there must be a strike. When a maturity date and an issue date are present, the issue date must be smaller than the maturity date.
- *Market data consistency.* In case of a quote with a Bid, a Mid and an Ask consistency can be checked. Bid must be smaller than Mid, which must be smaller than Ask, and Mid must be the average of the Bid and the Ask. For a quote at end of day of a trading session that contains the Open, High, Low, and Close fields, the Low must be the smallest and the High the largest fields.

- *Plausibility checks.* In this case price movements can be checked against historical standard deviations or versus the change in a benchmark.
- *Tolerance settings* on price behavior. Different data repair and data validation functions will be chosen with different parameters depending on the instrument type. A simple example is a threshold on return and marking everything that changes by more than a certain percentage or certain number of standard deviations. In the case of curves, we could use different percentage change tolerances for a curve; a different tolerance for the 1-week point versus that of the 10-year point. Stricter checks will be imposed on OECD currencies and there will be a higher tolerance for instruments where bigger swings are expected such as small caps, exotic currencies, and illiquid corporate bonds.

6.5 QUALITY METRICS: INFORMATION MANAGEMENT SUPPLY CHAIN KPIs

KPIs are a natural follow-on from treating the supply of quality data as a service. If the supply of quality data has evolved into a daily practice rather than a set of ad hoc activities, it makes sense to track it as a continuous operation. In this section we discuss quality metrics and KPIs. We are moving from metrics focused on the quality of the data to metrics that look at the quality of the process.

- *Data-driven KPIs.* These are lower-level KPIs constructed bottom-up from the systems and data. They will measure quality aspects such as those discussed earlier and can be used to gauge software and content services. They range from technical/system health information to information on data volumes.
- *Process-driven KPIs.* These are higher-level KPIs that reflect critical business metrics such as the number of missettlements and the percentage of client churn. These KPIs would be the basis for SLAs with services providers.

The closer the KPIs are to the actual cost base or correlated to revenue generation of the business, the more useful they are. Quality metrics and KPIs should drive efficient processes and help improve customer experience. Increasingly, analytics built on data directly or indirectly provided by customers will drive product development, client communication style, and marketing (see, e.g., King, 2010). For instance, any important event that can be inferred from public social media and/or transaction analysis could lead to alerts and client communication.

If we take core banking as an example, we can break up services by client segment as summarized in Table 6.2.

KPIs on information quality should be driven by KPIs on the services provided to these client segments in Table 6.2. Low-level KPIs can be generic and used for all client segments and would improve basic data, such as checks on the date of birth for retail clients and checks on their zip/postcodes. Certain

TABLE 6.2 Core Banking Services by Client Segment

Client segment/ service	Retail clients	Mass affluent	HNWI	SMEs	Wholesale/ institutional
Credit	Overdraft	Consumer financing	Larger-scale loans, special structures	Bank loans	Bonds issuance
Investment	Savings account	ETFs/ PRIIPS, funds	Structured products Wealth management, hedge funds	Treasury services, FX/ deposits	Bespoke investment vehicles, funds, OTC derivatives
Other services	Promotions, cobranding	Mortgages, college funds	Tax services, Estate planning	Corporate banking, international payment services	Transaction services, cash management

HNWI, High-Net-Worth Individual

properties are correlated with client behavior and can be used to build higher-level KPIs and indeed to build prediction models of client behavior.

KPIs built out from underlying basic data can be used for client services. Indications of life events, including children, marriage or divorce, a new job or job loss, or a move, used to be uncovered via account managers. Nowadays, automated transactions analysis and information volunteered by clients provides a scalable basis to achieve the same and includes actions following:

- stopped salary payments (this can indicate loss of job or illness);
- large transactions;
- idle cash (this can trigger investment proposals);
- changed spending patterns;
- new recurring income (this can indicate a new job or new business).

Because the information management problem is largely a logistic problem (getting the right information at the right place at the right time), metrics from the supply chain world provide a useful perspective. Concepts from supply chain management could not be applied just to the quality of the information but also on the success by which the instrument and transaction life cycle are managed and by which new content is produced by the institution. In supply chain management the focus is on logistics, distribution, and tracking inventory (balance sheet financial products and cash in the case of financial services). In Fig. 6.3 we discuss several supply chain management metrics.

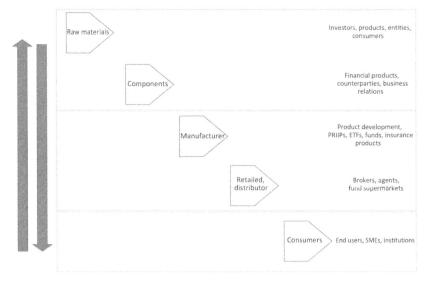

- Data flows down: information distribution, transfer of risk/title.
- Data flows up: more information providers, feedback, consumer needs, faster communication.

FIGURE 6.3 Supply chain perspective.

6.5.1 Throughput

Throughput refers to basic volume and can refer to the scalability of an information infrastructure. How many messages can be processed per second? How quickly can information be loaded and retrieved? What is the speed with which orders are filled?

6.5.2 Fill Rate

Fill rate refers to the percentage of customer orders that can be filled out of inventory immediately. In information management terms, it could also refer to the proportion of queries from users and downstream systems that can be answered immediately from a golden copy information inventory of security master, counterparty, or corporate actions data. The questions that cannot be answered (unfilled orders) result in a query to external parties. The costs of keeping an item on inventory and keeping it fresh have to be measured. Content costs can often be directly attributed; a portion of hardware, staff, and overhead cost need to be allocated on usage. This leads to requirements on data usage tracking. Often a financial institution is not fully aware of what content has been bought and of what information is available where. This means that there can be unused content inventory available.

TABLE 6.3 Balanced Scorecard

Financial aspects	Internal aspects
• Costs of inventory • Ad hoc costs • Savings/ROI	• Department budget • Planning/new content
Customer aspects	Training aspects
• Scoring on KPI metrics on supply chain management (SCM) • Customer satisfaction	• Knowledge level of data management department • Capability to take on new types of content, new products to process

KPI, Key Performance Indicator.

6.5.3   Balanced Scorecard

Balanced scorecard (the term was originally introduced by Robert S. Kaplan and David P. Norton in the early 1990s) refers to qualitative metrics to measure the efficiency of the information supply chain. An example of a balanced scorecard in the area of information management could look, for instance, like that shown in Table 6.3.

The balanced scorecard—as the name suggests—forces a much more comprehensive way of looking at a business function, balancing costs and benefits. It offers a high-level perspective that can sometimes be very refreshing for data supply chain practitioners who can otherwise get bogged down in detail.

6.5.4   Cycle Time

Cycle time is the time that elapses between a system and a user requesting a certain piece of information necessary for a process and receiving it. It is related to the *fill rate*. The target average cycle time will be set at different levels for different business processes. When you are trading or responding to a client query, turnaround needs to be fast. When answering to a query from a regulator or tax office, you normally do not need to be able to recover the information online. The required cycle time can vary from microseconds in quantitative trading strategies to hours or weeks when preparing financial statements. Table 6.4 gives some indication on business activity versus response times, from fast to slow.

The concept of *manufacturing cycle time* is the time spent between receiving the last bit of needed raw or untreated data from the content supplied and the delivery or update of the golden copy master information. Note that normally the infrastructure can also be set up so that the most frequently requested items will have the fastest cycle time.

TABLE 6.4 Cycle Times for Different Activities

High-frequency trading, arbitrage—microseconds	Turnaround time, relay quotes and orders to execution venues and confirm back
Pricing, predeal analysis—seconds	Output of quant models to price an OTC product for a client, predeal risk limit and capital cost check
Trade record completion—seconds	Lookup reference data for SSI, security master for trade confirmations
Intraday risk management—every few hours	Collateral repricing, checking limits versus market movements
Daily P&L calculation—hours	Preparation of daily EOD prices for P&L, exception handling, price verification
Curation of historical data sets and stress scenarios—days, typically monthly refresh	Historical data, correlations for risk metrics such as Expected Shortfall and Value at Risk, Exposure At Default (EAD), Loss Given Default (LGD), Probability of Default (PD) models for credit risk
Asset servicing—days to weeks	Corporate actions processing and client responses
Historical data lookup—days to weeks	Lookup past quotes/transactions for regulatory reasons for trade reconstruction
Financial year-end, acquisitions—weeks to months	Preparation of annual financial statements, due diligence on M&A

6.5.5 Defects Per Million Opportunities (DPMO)

This is a concept from the Six Sigma process improvement methodology and is defined as follows:

$$\frac{\text{Total number of defects}}{\text{Total number of opportunities for defects}} \times 1 \text{ million}$$

The 6 sigma goal is a 3.4 DPMO maximum (stemming from the background of Six Sigma: six standard deviations). The number of defects can be defined in an SLA: take a set of financial objects and a set of attributes to be tracked for those objects. Whenever the value of one of the fields delivered either on-request or as part of an agreed daily delivery turns out to be wrong at that time or delivered too late, this is a defect.

What is more difficult to measure is the number of opportunities for defects. Accounting for the number of disparate data formats, the lack of standards, the overall entropy, the number of sources, and the often lengthy supply chain means that for the financial information supply chain, the number of opportunities is very high. Let's illustrate this by making a ballpark estimate of opportunities. We take a typical architecture with representative numbers.

- seven ultimate sources;
- three redistributors or master data providers;
- three separate input channels in a bank;
- four security master data stores;
- five different export formats;
- sixteen downstream interpreting systems that keep copies of the data;
- one million instruments + 20,000 counterparties + 200,000 corporate actions, a total of 1.22 million records.

Assuming we update each bit of information once per day, this set of numbers would lead to 1.22 million $\times$ 16 $\times$ 5 $\times$ 4 $\times$ 3 $\times$ 3 $\times$ 7 or nearly 25 billion opportunities for a defect. Taking the Six Sigma goal of 3.4 DPMO would lead to no less than 83,623 defects in the master data. Adding transactional data to the master data example further complicates the story.

6.5.6   Perfect Order Measure

This is roughly the error-free rate at each stage of the order where we interpret an order in information management terms: an order is a query of a user or downstream application that will act on the data. Perfect order measure hence refers to the error-free rate at each stage of the supply chain. Note that it is often very difficult to measure the quality at each stage, partially because you do not have the complete information picture, and partially because various formats and identification standards cloud the picture. If you do have some of it in place, you can confront your external and internal sources with the score and report cards. The information on this Perfect Order Measure can be collected through information from the consolidation and validation process plus by feedback from the users on defects they report. (Fig. 6.4).

Note that there will be special "critical items" in the set of financial objects that you give extra special sourcing and validation treatment because you *cannot* afford to be wrong or off-market on these, for example, the EUR/USD exchange rate, the government bond curve, and blue chip equities. The specific critical items depend on the institution's business.

6.5.7   Inventory Turns

In Supply Chain Management, this is defined as the number of times inventory turns. In the context of data, we can relate this to data usage. How often is something *shipped* out of the master data database, and what portion of the sourcing/validation work is never used? This can be used in turn to drive internal cost allocation and to detect idle data.

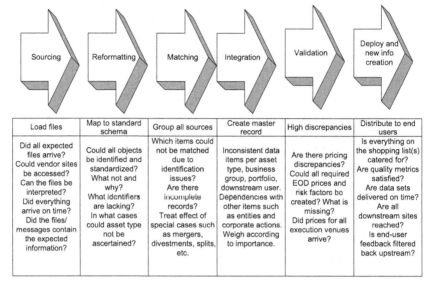

Sourcing	Reformatting	Matching	Integration	Validation	Deploy and new info creation
Load files	Map to standard schema	Group all sources	Create master record	High discrepancies	Distribute to end users
Did all expected files arrive? Could vendor sites be accessed? Can the files be interpreted? Did everything arrive on time? Did the files/ messages contain the expected information?	Could all objects be identified and standardized? What not and why? What identifiers are lacking? In what cases could asset type not be ascertained?	Which items could not be matched due to identification issues? Are there incomplete records? Treat effect of special cases such as mergers, divestments, splits, etc.	Inconsistent data items per asset type, business group, portfolio, downstream user. Dependencies with other items such as entities and corporate actions. Weigh according to importance.	Are there pricing discrepancies? Could all required EOD prices and risk factors be created? What is missing? Did prices for all execution venues arrive?	Is everything on the shopping list(s) catered for? Are quality metrics satisfied? Are data sets delivered on time? Are all downstream sites reached? Is end-user feedback filtered back upstream?

FIGURE 6.4   Information supply chain with quality questions asked at each stage.

6.5.8   COPQ

An interesting metric is the *cost of poor quality* (COPQ) of information. These are the costs that would vanish if systems, processes, and data content were perfect. Making the COPQ explicit is critical as this is the main number to be estimated for any business case for quality improvement projects. Direct costs on staff, redundant repositories, and content are relatively easy to measure. If opportunity costs are woven into the equation, the number would become very high. We list different ways to assess business impacts in the next section on ROI to help think about which costs are avoidable and which are not.

6.5.9   Other Measures

There are many other measures that could be used for metrics in SLAs. These include *performance to promise* that measures adherence to the terms of the SLA. *Material value add* is the measure of value added in the process and is normally expressed as [(sell price) − (material cost)]/(material cost).

Looking at the ratio of total cost over the information supply chain divided by content costs, we will see content cost represents a small fraction. Even if we leave out the costs of actually uploading content in applications, the amount spent within an institution to treat the content is higher than the price paid for the content. We can interpret these *holding costs* as the costs of the data distribution infrastructure including interfaces and mappings: the cost of putting a piece of content through the supply chain. In information terms, we also need to check how much history is kept online and what level of audit is kept on the information. As discussed in Chapter 4, ad hoc, historical requests for risk, audit, regulators, and clients have increased.

6.5.10   KPIs and Root-Cause Analysis

The trend is to measure the cost of data and the knock-on cost of poor data. Data inventory tools [market data inventory tool providers include Screen Infomatch (http://www.screeninfomatch.com), MDSL (https://www.mdsl.com/), and The Roberts Group (https://www.trgrp.com/)] can keep track of the set of data assets and who uses what but need to be complemented by specific metrics on the data quality taking user feedback into account. KPIs often come with different levels of detail, from summary information displayed on corporate dashboards to detailed information needed to improve the situation (Fig. 6.5).

If you closely study KPIs, you may also be able to deduct the potential for process failure. This means you will be one step ahead and through root-cause analysis prevent losses. Failed processes will lead to losses and KPI dashboards can provide early warning. The process of putting together these dashboards and having a discussion between different stakeholders about which KPIs are the most relevant will already provide insight into where processes are brittle and can lead to improvements. Historical data on the metrics needs to be tracked to spot trends and to be able to deduce performance and costs balance. Supply chain management is about the balance between costs and quality. It will perhaps be possible to reach 100% accuracy of information, but at some point further improvement might not be economically feasible.

Clearly defining the relevant metrics and knowing how to act on deviations from target levels is critical. Relevant statistical indicators will be applied to KPIs and alerts will be set based on thresholds. Setting thresholds to identify the unexpected is the basis for trend and operational risk analysis. Before that it is important to explicitly state what outcomes are expected taking into account information such as:

- expected periodic variation, which can include seasonal effects for energy prices and commodities, specific monthly, weekly, or daily rhythms in, for instance, trade flows and annual or quarterly cycle of dividends;
- the "normal" range or bandwidth of a measurement to be able to distinguish background noise from real outliers;
- the historical high/low cases to point out when something unprecedented is happening;
- expectations on correlations between several metrics and which ones are expected to move in lockstep. These are expected on whether certain indicators should be leading or lagging other indicators. This can point out when the assumptions about logical relations between different indicators need to be revised.

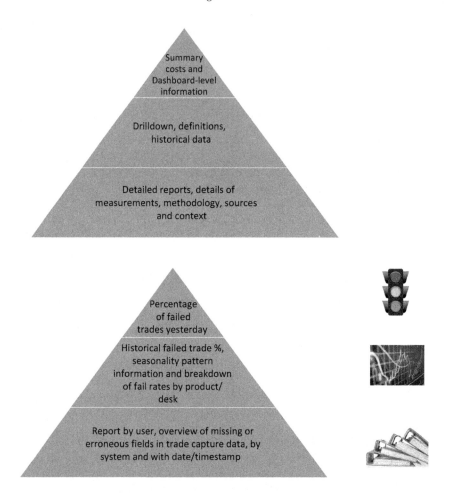

FIGURE 6.5   KPI detail levels and example for STP rate. *KPI*, Key Performance Indicator.

Standard statistics would measure deviations from the mean, the number of standard deviations from the mean or historical correlation, or volatility. They could test against certain hypotheses and provide confidence intervals about the probability distribution of events. Statistics can be set on both low-level, data-driven KPIs and also on high-level, process-driven KPIs. KPIs also measure the number of disruptions in the information supply chain, for example, when we have unknown instruments, unknown customers or counterparties, or unknown settlement locations, or when there are no prices for an instrument. This information has to be fed back to the information supply chain infrastructure and perhaps down to the content supplier if that is where the break occurred and to fix the gaps or errors in the data.

Examples include:

- counts on costs, defects, and number of instruments;
- response times in requesting new instrument setup;
- errors in counterparty setup;
- settlement delays or reversals;
- fill rates of orders;
- responsiveness of counterparty when additional information is requested;
- the tracking error with regard to an index;
- deviations in cost per transaction;
- credit quality of a loan portfolio portion through portion of delinquent loans or the amount of loan write-offs;
- wrong valuations;
- collateral quality and composition;
- timeliness and accuracy of corporate event information.

6.5.11   Defining and Monitoring KPIs and Their Use in an SLA

Different departments in a firm worry about different quality measures. Traders need information around price behavior, drivers, and arbitrage margins, custodians need clear identification, and timely details on events such as dividends, investors need to know about controls and valuation procedures, regulators about transparency, investor protection and capital adequacy and back-offices about smooth operations and low cost per transaction.

KPIs are set at various points in the trade life cycle and the corresponding supply chain. They will be of interest not only to the firm's management but also to analysts that want to compare different companies and to regulatory agencies who want to check the soundness of control procedures. In Fig. 6.6 we list various information interests at different stages of the transaction life cycle.

Example KPIs for primary aspects on a data service are provided in Table 6.5.

KPIs with target levels and possibly penalty clauses are often the building blocks for a SLA. The other part of an SLA is an incident priorities table where different severity levels are defined with their target response times. An incident priorities table could include definitions of the response and resolution times:

- The timing to acknowledge an incident for the client will be *xx* min (regardless of the severity).
- Time to respond begins when an issue is acknowledged by the data service provider and ends at the time the data service provider provides confirmation that it accepts the issues, understands the issue, and has identified and coordinated a response to remediate the issue.
- Problem Resolution begins when the Response Time is provided and ends at the time the data service provider completes actual problem resolution.

The time to respond to and resolve the incident can vary according to the classifications in Table 6.6.

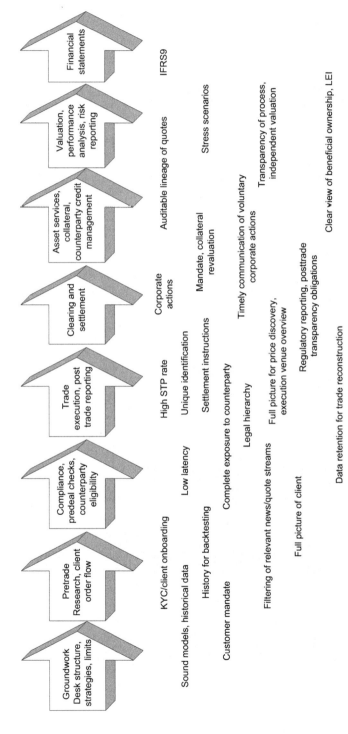

FIGURE 6.6 Different information interests at different stages of the transaction life cycle.

TABLE 6.5 KPI Examples

Service type	Service measure	Performance target	Minimum performance (%)	Measurement interval	Weight (%)
Availability	Availability of FTP site for live data	FTP site availability	x	Monthly	x
Delivery timeliness	Delivery of data on agreed schedules	File deliveries on schedule	x	Monthly	x
Incident handling	Time to handle	Handle incidents within timings defined in Incident Handling Matrix including escalations	x	Monthly	x
Request handling	Time to handle	Handle requests within timings defined in Request Handling Matrix	x	Monthly	x
Quality	Data accuracy rate	Deliver correct data attributes per requirements and business rules	x	Weekly	x

KPI, Key Performance Indicator.

6.5.12 KPI Best Practices

Proper KPIs combine a rule for measurement, an expectation of the outcome of this measurement, and a decision tree on how to treat deviations against that expectation (Fig. 6.7).

Best practices for selecting and setting KPIs include:

- KPIs should be *comparable* to those of your peers or other departments to see where you can improve and where you can improve most for a minimum effort. If multiple companies share a common KPI methodology and pool

TABLE 6.6 Incident Response Times

Code	Name	Definition	Response time (min)	Resolution time
1	Critical	• Service/Application down for all users or several systematically significant users • May lead to a major financial impact and/or a major damage of the company reputation • No workaround is available	x	x
2	Severe	• Service/Application down for several significant users or one systematic user • Service/Application degraded for all users or systematic users • May lead to an important financial impact and/or a major damage of the company reputation • No workaround available • UAT environment not available	x	x
3	Significant	• Service/Application down for at least one standard user • Service/Application degraded for one or several standard users, Clients • May lead to a significant impact • Workaround readily available • UAT environment degraded	x	x
4	Low	• Minor problem but the resolution or intervention will be planned with the user(s) (no or insignificant impact for the company) • May lead to a low impact	x	x

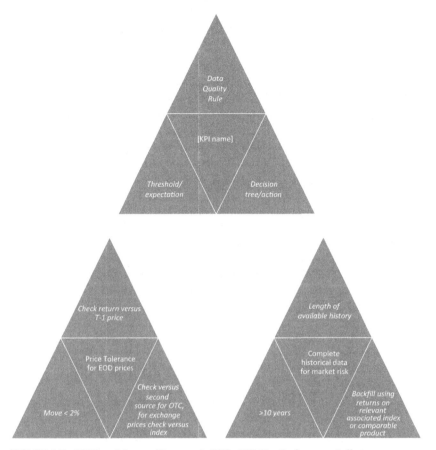

FIGURE 6.7   KPI breakdown with example KPIs. *KPI*, Key Performance Indicator.

data, this can help them see how they do relative to their peers. This also holds for benchmarking maturity levels in data management. This benchmarking can be done *internally* from department to department or externally with a peer group. Regulators too would benefit from using the same metrics across all the firms.

- Take great care in choosing the right indicators that reflect the operational efficiency of *your business* rather than copying a set of indicators from another firm. KPIs are like any other statistic: you want to condense information as efficiently as possible without losing the big picture and the main trends. In data terms, if you measure parts of the information supply chain, you would expect the KPIs to reflect the important *quality* aspects discussed earlier.

- The selected KPIs should reflect the *purpose* of the specific relationship (in case of outsourcing to a third party) or the purpose of the measurement. The purpose could be customer satisfaction—through metrics such as the response time for ad hoc queries or the percentage of portfolios adequately serviced with pricing/models, for example, 100% accurate NAV, nothing contested, number of mispricings, STP rate for settlement; the purpose could be cost (USD/transaction); it could be the support of higher volumes through metrics such as the uptime of systems, for example, 5 × 9 availability (99.999% uptime or about 5 min maximum unscheduled downtime per year), number of new instruments set up, number of corporate events processed, number of portfolios serviced, latency, throughput in messages/second, and so on. Similarly, it is important to understand what the result of improving on the current values of the KPIs will be. Will this be happier clients? More revenue? Lower costs? All of the above?

- Do not collect *too many* KRIs. KPIs can be strongly correlated so increasing the number of KPIs does not necessarily provide additional information. Furthermore, KPIs should be easily collected. If they are costly or time-consuming to collect, this will not (always) happen and you will have either an incomplete on an imprecise picture because people will rush to deliver something and interpret the data to be delivered in different ways. Reports with hundreds of KPIs will not be acted upon in an appropriate and efficient way and will probably not even be read: people will not be able to see the forest for the trees.

- Make the definition of the KPIs *consistent* among the various processes that need to contribute them. Inconsistent KPIs can never be rolled up into aggregate information or be the basis for operational risk charges allocation to business divisions. Consistency can also help the operational risk reporting and will definitely benefit the regulators that need to interpret and compare these reports.

- The KPIs should be under the *control* of the responsible party or of the party to which this process is outsourced. For example, if an external party that manages the data is measured on lead time in setting up new instruments or customers in the systems, it could be that lead time heavily depends on the financial institution's own risk control procedures.

- KPIs should *support root-cause analysis* and predict trends.

- KPIs should have *owners* that are responsible to reach a certain value with the KPIs. There should be a feedback loop back to responsible people who can *act* on the KPIs.

- KPIs should be *process* driven rather than *data* driven. Through process-data mappings, the underlying data elements that determine the success of the products need to be identified.

6.6   EXPOSING AND CONTROLLING INFORMATION UNCERTAINTY

In information theory, *entropy* refers to the level of surprise or the amount of uncertainty in information. In a financial infrastructure context, we can apply the concept as a measure of the efficiency of the information supply chain.

The different steps in a supply chain before it reaches a place where an end user or system actually *acts* on it combined with the number of different departments and locations provide an indication of data quality and hence of the trust that users can have in it.

Information supply chains can be like a game of Chinese whispers with increasing uncertainty on the accuracy of the information as the number of places where it is touched grows. Regulatory demands drive new vendor content and software products that need to be integrated with existing infrastructures. If we look at firms' data architecture from a bird's eye perspective, they look like a clogged-up river delta. Data flows branch out and come together. Some channels run smoothly from source to destination but others are clogged up with silt causing backlogs and flow to come to a standstill.

Improvised Excel sheets and other EUDA are used to intervene for time-to-market reasons but typically last longer than anticipated and become part of the infrastructure. The information system, spreadsheet, and database jungle is also the result of past mergers and acquisitions and recent consolidation. The stratigraphy of acquired companies can be observed in the information infrastructure. The reason that forensic accountants and EDP auditors have to act occasionally as amateur archaeologists is that it is hard and costly to get the information supply chain perfect. Users will always be reluctant to cede control and surrender the spreadsheets.

The extent to which information is dissipated around an institution equals the effort it takes to create a coherent and comprehensive picture of the state of the business. Factors that contribute to the overall information entropy in an institution and that would need to be assessed in any initial data quality diagnosis include:

- the number of independent data streams coming in;
- the volume of transactions and the number of different types;
- the range of products traded;
- the dispersion of users across geographies and time zones;
- the number of independent data stores and how often master data is stored inside business applications;
- the number of intermediary actors and steps in process flows;
- the number of definitions of data fields;
- the number of independent entry points (via manual entry, client submission, commercial data sources) of the same or overlapping data;
- the prevalence of Excel and other EUDA in operational workflows;

- clarity of the ownership and change management per data domain;
- the number of people having write access to the same data, in which departments;
- the dispersion in human languages, units of measurement, and file formats in the firm.

The higher the entropy, the higher the overall complexity, cost, inertia, and inability to cope with new information management requirements. Through looking at the different elements that went into creating it, we can investigate the potential for improvement. Operating a security master twice does not *double* the work; it *triples* the work because you also need to reconcile between your two internal data stores. New data demands in a poor infrastructure lead to additional staff doing reconciliations and rework.

Mitigating factors that can reduce the data uncertainty and improve the efficiency of information lifecycle management include:

- use of single data standards to solve the incompatibility and interpretation issues;
- shared services and avoiding duplicate sourcing, processing, and storing of the same information;
- data governance via clear ownership and change management procedures per data domain;
- controls and operational risk reporting framework;
- insurance against major operational risk events;
- metrics and dashboards to track quality KPIs and spot trends;
- a feedback mechanism to improve data quality internally and of data coming in via customers and suppliers.

How to go about implementing these mitigating factors depends on whether the firm is organized by geography, client segment, or product or via a matrix organization and whether there are cross-vertical shared services that can drive standards and common services (Fig. 6.8).

Even the best information architecture designed and set up from scratch will be a reflection of the business requirements *at that moment*. Because of changing needs, it will be tweaked, abused, and retrofitted. Even in the case of a sound underlying architecture, shortcuts will be taken. This is just one reason why *metrics* and periodic benchmarking against industry best practices and maturity models are important: to keep everyone on the right path.

The second law of thermodynamics seems to apply to data infrastructure as well. Without intervention and controls, users will take shortcuts to achieve short-term goals. Without a proper data management practice, information uncertainty or entropy will increase. The only way to stem and reverse the pull toward data degradation is to invest energy into the system.

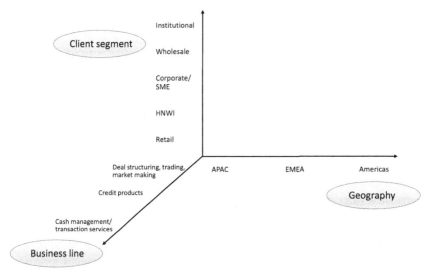

FIGURE 6.8   **Organizational dimensions of financial institutions.** *HNWI*, High-Net-Worth Individual.

6.7   QUALITY AUGMENTATION AND REMEDIATION PROCESSES: WHAT TO DO WITH KPIs?

After defining and measuring KPIs, they should be exposed and lead to corrective actions. We will define the terminology of Business Process Management (BPM) and Business Activity Monitoring (BAM) and its relevance to financial process metrics.

Dashboards present a collated view of KPIs, and glue different steps in a workflow together to provide overall insight into the performance and health of a process. "BPM" refers to the control and transparency of various steps in a business process and is at the crossroads of the fields of management and information technology. It addresses the process aspects of enterprise architectures. "BAM" refers to the definition of certain metrics that need to be maintained and conditions for alerts and actions. Compared to dashboards, BAM are more real time and driven by events and oriented around business processes. *Business intelligence* ("BI") refers to historical analysis and data mining to uncover trends that BAM would expose in real time. Dashboards, BAM, and BI are all techniques to gain insight into (how to compare) processes and require underlying KPIs. BAM could, for instance, be used as inventory management in cases where proactive alerts can be created such as:

- limiting breaching or approaching limits;
- cash balances in accounts, real-time inventory of specific instruments;

- collateral portfolio quality, for example, when certain credit or industry concentration limits are reached, triggering collateral top-up or substitute calls;
- market-making inventory, leading to adjustments in quotes and size quoted or direct buy/sell orders to change the inventory level;
- margin calls, triggering cash transfers;
- checking outstanding corporate events where deadline approaches and sending alerts to account holders to inform them to make a choice.

BI could be used in a more postfact reporting and analysis environment, for example, on STP rates to investigate which accounts and which products cause the largest number of breaks to subsequently invest in the data quality of those areas. More powerful BI tools can be used by data scientists or business engineers; new roles that combine business knowledge with broad IT skills.

Initially, BPM and BAM will show only the symptoms, but if well set up, they should lead to diagnosis. Diagnosis should lead to prescription. Prescriptions should lead to cure. The best practices for KPIs discussed earlier should mitigate these risks. Note however that logic and value in the tools used can never be a substitute for intelligence needed in configuring and using them.

Dashboards and corrective actions on KPIs are the logical conclusion of collecting quality aspects of the information supply chain. The process starts with identifying which content is required and what the required service levels are for the processes that depend on that content. Setting up KPIs is the next step. The relative weights of each of the quality dimensions will vary by business function. Data quality elements can represent competitive advantage; it may not be the goal to reach a certain *absolute* level of quality but you may merely need to be *relatively* better than everybody else. If that is the case, you will have a competitive advantage even with far from perfect quality data.

Operational KPIs and metrics from predictive analytics are closely connected with manual intervention:

- When missing data is flagged: go and find it internally or go and buy it from an external on-demand data service, or contact a customer to supply missing personal data.
- Suspect data: go and find corroboration from another source.
- Make an automated call; send an email or texts. Data collection portals can be set up to come to more dynamic data collection and data management processes where users can confirm or complete data themselves.

SLAs define the expectations between two parties for a service provider relationship. They define what is promised by the service provider, how it will be measured, how it will be priced, and what the effect of deviations from the service is. KPIs are a useful tool to have clear and concise agreement on measures in place and to summarize the expectations. Since more and more services are provided either by common services departments within institutions or by third parties, it is important to set the expectations right from the start.

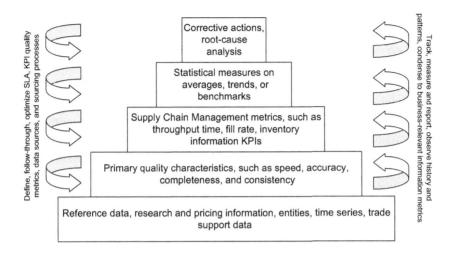

FIGURE 6.9   **Data metrics reporting and feedback loop.** *KPI*, Key Performance Indicator.

KPIs as broken down in Fig. 6.7 should not be used just as a reflection of the status quo. They present an opportunity to drill down into the root causes in order to improve the processes. Frequent feedback from observed values in KPIs reinforces the scrutiny of processes and leads to improvements in the sourcing, processing, and distribution of information. The feedback on quality needs to travel the opposite direction from the direction of the information supply chain. Without this, there will never be any improvement. The feedback loop can look like as shown in Fig. 6.9.

Trends we see in KPIs, include more granular costing on data, tighter controls on ownership, and the increased amount of contextual data kept. Future developments, include metadata and confidence scores. Every piece of data may get its Facebook profile tracking where it goes, what it does, what it is linked to and feedback from users. Also, there will be KPIs more closely linked to business processes. We will go to a situation where every action or transaction is costed out in a more granular way. Management/financial accounting will pervade operations. Data too will be charged for in a granular fashion and content license agreements will accommodate or even stimulate this since content owners like to have a full view of the use of their data.

6.8   THE ROLE OF DATA STANDARDS

We mentioned some of the data standards in Chapter 3. Standards exist at different levels:

- at the individual field domain level: a standard set of country of currency codes;

- at the identification level: a standard to identify legal entities or financial instruments (ISIN code is ISO standard 6166; LEI code is ISO standard 17442; see also https://www.leiroc.org/);
- at the level of transactions;
- at the level of complete models of the financial services world, typically split up into different subdomains.

Standards put users in different departments in different firms at different steps of a business process on the same footing. Controlled vocabularies with unambiguous definitions and clarity about the relations between different terms in those vocabularies are a prerequisite to efficient automation: automating ambiguity typically is a recipe for disaster.

Standards can come out of the ISO standard-making process (a good overview of this process can be found on http://www.iso.org/iso/home/standards_development.htm) that organizes technical committees with industry representatives and includes appointing a Registration Authority that has to facilitate implementation of the standard. For example, SWIFT is the Registration Authority for standards including ISO 20022 and the Legal Entity Identifier. Standards can also be driven directly by the industry, as was the case for FIX and FpML. Common standards include those given in Table 6.7.

In IT, the word *ontology* refers to a formal naming and definition of types, properties, and relationships of the entities in a particular domain. There can be ontologies, for instance, for currency options trading, for corporate actions processing, or for credit risk management.

TABLE 6.7 Transaction Data Standards

Standard	Scope	Description
FpML	Transactions, focused on derivatives	Development of standard integrated with ISDA
FIX	Transactions, mostly listed products	Developed by Fix Protocol Limited, a nonprofit industry association
ISO 20022	All financial messages	Cross-border financial communication. SWIFT is the registration authority of the standard
XBRL	Exchanging business information, mostly financial statements	Developed by XBRL International. Strong connection with financial accounting. Widely used for reporting of financial statements and corporate tax reporting

Further information on the respective standards can be found on http://www.fpml.org/, http://www.fixtradingcommunity.org/, https://www.iso20022.org/faq.page, and https://www.xbrl.org/, respectively.

The purpose of the ontology is to get common meaning to come to a common human understanding of terms and relations. In turn, this allows for a common machine understanding, and a more effective automation and information exchange between applications. Ontologies define all variables that could play a role in the automation of certain business processes and set the potential relations between them. Establishing fixed, controlled vocabularies with their relations allows not just automation but also automated reasoning.

Ontologies not only describe the meaning of terms; they also use inferences to provide data classification, validation, and links. From the definitions, relations between different objects can be inferred. Ontologies typically supplement existing technologies by adding clear meaning and links. Ontologies and data standards generally help with:

- Lower software development and maintenance costs and faster time to market when building new processes on the same ontology domains.
- A shared understanding through a shared vocabulary. A common data model for operational processes (trade processing, security services, tax services, fund administration) should lower costs and increase timely customer access to information.
- Providing a consistent basis on which to define KPIs and other metrics. This can be used for comparing and benchmarking quality when breaking up business processes across different firms when outsourcing.
- Support for financial and systemic risk intelligence. Regulators especially need standards as they are superaggregators of financial information.
- Regulatory compliance, KYC but especially BCBS239 that prescribes requirements for risk data aggregation. Any rollup of information requires common standards. You cannot add numbers if you do not know in which base they are represented.
- Start of industry alignment on common meaning of financial concepts, the basis for shared industry services and interfirm interoperability.

Ontologies are the basis for inferences and decision making and can lead to *smart contracts* that have built-in rules to prevent errors or to trigger actions. Smart contracts are essentially automated protocols that facilitate, verify, and enforce the terms of a financial contract and kick off the required execution workflows. Compared to current IT processes, the data model and associated processes are more closely coupled.

Smart contracts can be combined with distributed ledger technologies to automatically transfer assets based on certain boundary conditions being fulfilled.

In Table 6.8 we compare the current state of data modeling and automation with the potential in smart contracts.

Gruber (1995) introduced ontology into computer science in the 1990s. Gruber introduced the term to mean a specification of a conceptualization: "An ontology is a description (like a formal specification of a program) of the concepts and relationships that can formally exist for an agent or a community of agents."

TABLE 6.8 Smart Contracts Versus Current State Differences

Smart contracts	Current state
Transparent, low or no enforcement cost	Opaque, potentially high enforcement cost
Composed of clear building blocks and actions outlining, for instance, a counterparty, a transaction, a collateral call or a payment	Complex because only loosely integrated, human interaction, or IT glue in the form of spreadsheets and point-to-point interfaces needed
Trusted, assets can be moved automatically	Error-prone, subject to fraud
Verifiable, terms shared between parties	Imprecise, each party has their own internal IT infrastructure and data model
Semantic, meaning embedded in the terms and conditions	No meaning or meaning has to be gleaned by human intervention or custom IT applications

The financial crisis has led to much more concrete interest in standards:

- Tighter margins have led to higher scrutiny on cost and interest in the efficiencies standards can bring.
- Customers expect online price discovery, transactions, and reporting. Robust technology infrastructure behind those functions would not be possible without standards.
- Regulatory reform has led to more standardization in business processes and commoditization of certain products, for instance, central clearing and post-trade transparency for derivatives.
- The industry is cognizant of the dangers and financial ramifications of ambiguity in transactions. Counterparties of interest rate swap or currency swap trades will not accept ambiguous phrasing and demand transparency. Lack of standards can also lead to internal confusion on terminology.

The EDM Council is driving the promising development of the Financial Industry Business Ontology (FIBO; see www.edmcouncil.org, trademark EDM Council). This is meant to be an open industry "common language" standard for defining the terms, facts, and relationships associated with financial contracts. FIBO covers financial instruments (product reference data), market data pricing, and financial processes.

Standards provide a platform on which to innovate—a foundation on which to build analytics. They help provide a common data foundation that can be used by the industry to pool data and to come up with shared services. Standards reduce the friction in handover points when data moves between applications, from data sources into applications, or from applications into reports.

Standards are not an esthetic end in itself but should be an enabler. The goal is not to standardize for the sake of it; the goal is to come to shared vocabulary to prevent avoidable cost and risk in data management. History shows that standards that come out of clearly delineated domains (corporate actions, financial accounting, derivatives, or securities transactions) have a much higher chance of adoption than standards that take on the world.

6.9   DATA MANAGEMENT MATURITY MODELS

Analogous to the capability maturity model in software engineering, maturity models have been developed to gauge the maturity of data management practices in institutions. Data is following IT in maturity models; the tools need to be better connected and data standards are a precondition for that. A very useful model is that developed by the EDM Council called "Data Management Capability Assessment Model" ("DCAM"; see www.edmcouncil.org for more information).

The DCAM defines the scope of capabilities that is needed for a mature and well-controlled data management practice within an organization. It was created through the EDM Council with industry participation and serves as a good benchmark for any firm assessing their capabilities or starting a new data management program. It is used as a checklist for objectives and deliverables, introduces a common vocabulary, and is also used as a basis to survey the industry.

The capabilities covered for scoring are organized into eight components:

- data management strategy
- data management business case and funding model
- data management program
- data governance
- data architecture
- technology architecture
- data quality program
- data control environment

Each of the components is split up into capabilities and those in turn are split into subcapabilities, for example, component 5 on Data Architecture looks as given in Table 6.9 (see www.edmcouncil.org for the full set of capabilities and subcapabilities; with thanks to Mike Atkin of the EDM Council).

The EDM Council has developed a scoring methodology to assess the maturity level in each of these capabilities. This is done on the following scale:

1. Not Initiated
2. In Process (Conceptual)
3. In Process (Developmental)
4. In Process (Defined)
5. Capability Achieved

TABLE 6.9 DCAM Data Architecture Component Example

5.0 DATA ARCHITECTURE	5.1. Identify the data	5.1.1. Logical domains of data have been identified, documented and inventoried
		5.1.2. Underlying physical repositories of data have been identified, documented and inventoried
	5.2. Define the data	5.2.1. Conceptual models are defined (ontologies)
		5.2.2. Logical models are developed (taxonomies)
		5.2.3. Attribute level "business" definitions are defined, documented and approved by relevant stakeholders
		5.2.4. Metadata is defined
	5.3. Govern the data	5.3.1. Data architecture governance procedures are established to ensure authorized as well as controlled use of data
		5.3.2. Data architecture governance procedures are in place and aligned with business governance processes
		5.3.3. Data architecture governance procedures are in place and aligned with technology

DCAM, Data Management Capability Assessment Model.

For example, for some of the capabilities listed in the previous data architecture example this scoring would look as shown in Table 6.10 for the beginning of Section 5.0 on Data Architecture.

The EDM Council runs regular benchmarking exercises (see EDM Council, 2015). Many of the capabilities identified in the DCAM directly correspond with capabilities needed to live up to regulatory initiatives such as BCBS239 on risk data aggregation. An EDM Council recent survey showed industry scores on individual capabilities averaging between 2.66 and 3.53. At the lower end of the scale there are capabilities on data profiling, lineage and metrics. At the high end of the scale there are data management strategy, alignment, and funding. From this it seems the industry knows it needs to do something but the strategy hasn't yet filtered down to translate to an end-to-end transparent sourcing process.

TABLE 6.10 DCAM Data Architecture Capabilities Example

	Not initiated	In process (conceptual)	In process (developmental)	In process (defined)	Capability achieved
5.1.1. Logical domains of data have been identified, documented and inventoried	Logical data domains have not been defined	Logical data domains are proposed. Business stakeholders are identified to participate in the identification process	Business stakeholders are selected to identify logical data domains. Business stakeholders are confirmed, engaged and participate	Logical data domains are validated by involved stakeholders	Logical data domains have been identified, prioritized and sanctioned
5.1.2. Underlying physical repositories of data have been identified, documented and inventoried	Physical repositories have not been identified	The identification of physical data domains is underway	The inventory of physical domains is identified and shared with involved stakeholders	Physical repositories are linked to logical domains. The linkage has been documented and verified	Physical repositories have been designated and the inventory is actively maintained

DCAM, Data Management Capability Assessment Model.

6.10   ROI OF DATA MANAGEMENT PROCESSES AND QUALITY MANAGEMENT

Inefficient data management processes impose both direct and indirect costs, direct because the redundant sourcing, storing, and processing cost money and errors in data lead to rework to fix them. More important are the indirect costs: redundant processing introduces uncertainty and lack of trust. Ultimately, data quality issues have consequences on a firm's relations with its customers, counterparties, investors, and regulators. An overview of cost types can be found in Table 6.11 (drawing in McGilvray, 2008, pp. 190–191).

There are different business impact techniques that can help proof value and prioritize data quality improvement efforts. Improvement efforts can range from the small-scale changing procedures, tracking basic data metrics and discontinuing duplicate storage, to large-scale, enterprise-wide efforts. It is best to start small to test what works and then see how proven small-scale interventions can scale to the organization. Impact techniques include (drawing in McGilvray, 2008, p. 163ff):

TABLE 6.11 Cost Types Overview

Impact type	Examples
Hard impacts—easily measurable	Customer attrition: loss of business due to clients switching banks, switching funds Regulatory fines due to misselling, fraud
Soft impacts—evident but hard to measure	Delayed decisions through perceived need to double check information Manual rework, quality checks by business users and risk and finance departments that feel they cannot trust the master data systems, local databases Leading to inertia, lack of competitiveness, and conflicts over data ownership
Decreased revenue	Opportunity cost because of long time to market for product development Customer dissatisfaction when getting incorrect/slow information or generally poor service in a mobile world Poor cost/volume KPIs, lower brand value
Increased cost	Organizational inertia, high cost due to duplicate data sourcing and warehousing, additional heads in operations and IT, workarounds
Increased risk	Regulatory intervention and internal audit due to brittle IT infrastructure and noncompliance with BCBS239 principles Opportunities for fraud in poor data management and data governance; opportunities to hide or abuse when risk, valuation, position keeping take place in end user–controlled applications
Lowered confidence	Regulatory audits, scrutiny due to high-profile data issues, fraud or trading scandals covered in press, loss of investor confidence and can lead to regulatory intervention in capital planning Poor management information systems

- *Anecdotal.* Most users would have stories on the effects of poor quality on orders, client satisfaction, or how it impacts their own daily work (e.g., redoing the same work).
- *Usage.* Comparing current and future new uses of the data and every process that will touch it. Usage of customer or product data in new regulatory reports can highlight quality issues and trigger improvement projects.
- *Root-cause analysis* by interviewing business users and repeatedly asking "why." As a rule of thumb, after five times you get to the real business

impact. This can help tracing data through its life cycle through the business processes. It is also often a good way to identify dead-end data flows and stale data pools.

- *Benefits/cost matrix* to qualitatively inventory the effects on an operation of poor quality but also to show how processes could be improved if the required data was more quickly available, consistent, or richer.
- *Prioritization* of the impact of missing and incorrect data by ranking different quality issues with their impact on specific business processes.
- *Process impact.* Show the effects of poor quality data on business processes. These would often be workarounds or heavy use of End User–Developed Applications such as Excel to overcome infrastructural data issues. Also, often extra headcount hides the fact that there are data quality issues.
- Gathering a number of *examples*on the impact of the low-quality data in terms of lost business, missed revenue opportunities, waste, rework, or regulatory fines is the best way to proof the case for a structural improvement. A number of anecdotes and KPIs on data quality can suffice if these can be extrapolated to a larger operation.
- To state a *return on investment*, a complete assessment is needed on the financial impact of low-quality data, on the investment needed in structural improvement, and on the quality level reached in the new state. Costs to keep the data quality after the project has ended need to be included too. It is easy to be too optimistic but data quality projects rarely achieve 100% accuracy—and even more rarely manage to completely prevent decay afterwards.

Larger-scale data quality improvement projects can include the use of some of the tools discussed in Chapter 5. These can be data governance tools to keep track of definitions, internal social media to collect feedback, and data cleansing tools. Specific data cleansing tools can parse existing data and find duplicates or gaps—especially when the domain and definitions are clear and there are many conventions that can be coded in these cleansing tools, such as in customer records and address information. Other tools self-discover and document processes and data flows—provided you let scanning and discovery software loose on your network. Tools need able hands to wield them. When there is no tool, custom scripts will be developed and mass database updates have to be done. Care has to be taken that existing operations are not disrupted. We will discuss the organizational dynamics of data management in the next chapter.

Business impact can be expressed in different ways (see McGilvray, 2008, p. 166ff) including:

- increased revenue;
- direct cost savings;
- operational efficiency through increase in KPI (e.g., lower turnaround time);
- increased employee productivity (turnover per employee, orders processed per employee per hour);

- reduction in risk in the sense of unexpected variation in outcomes, that is, more predicable processing time.

Relevant KPIs are needed to support the business case, for example, track number of manual interventions to proxy cost. Every manual intervention could cost 5 min; every approval could cost 2 min. Having this summary information will help make the business case. Track the number of occasions that led to external communication: how often was a client or counterparty contacted about an error in settlement, in valuation, or in reporting? Together with a database on operational losses, this could be used to gauge direct costs.

For financial services, complying with regulation is the cost of being in business. Data quality issues often become acute triggered by ad hoc reporting needs that cut across silos, in the financial crisis, for example. For chronic problems people will develop workarounds in the form of extra headcount and EUDA.

Regulation is a major driver of data improvement projects but improvement projects should not be sold internally purely on regulatory grounds. It is not so much "because you have to" but far better to present the revenue-generating opportunities and take the "because you want to" angle.

Data management switching costs are very high. Master databases and distribution tooling are sticky, precisely because of the entropy and the number of point-to-point connections. This has meant that organizations are often locked in for substantial periods with solutions that they are not really happy about. Conversely, vendors may not be sufficiently incentivized to provide world-class service since their clients have little choice. Therefore, when setting up projects to smoothen and reorganize the information supply chain, make sure to always address at least one immediate business.

Some predictions as to the future challenges of information management are easy: more volume, lower latency, more regulation, and larger geographical spread between different links in the transaction life cycle. If data quality is a comparative advantage, it is also a moving target as the average quality and thus the benchmark shifts and the bar is raised continuously. One constant is that risk will always end up where you least expect it. This holds for financial risk as much as for information management risk.

The higher the information entropy, the higher the potential ROI when you clean up and streamline your "content act." Reducing the entropy means reducing the opportunity for bad data and reducing the opportunities for data defects. This will mean lower costs in retrieving the information, higher productivity, and an easier path to compliance.

6.11   CONCLUSIONS AND FUTURE OUTLOOK

Information becomes useful only when it is acted upon. Presentation quality and easy accessibility of the data will determine its usefulness. Before it ends up in a decision-maker's hands, information goes through an often lengthy supply chain. Knowledge about this information supply chain, about changes and additions to information at each step, and about what happens at handover points is a condition of success for any activity that depends on this information. All too often information arrives too late, is incomplete, or is degraded at the place where it is needed.

A good understanding about what aspects of data quality matter to the users and how to measure it and come up with the appropriate KPIs is critical. Standards and maturity models can help here and are necessary preconditions and practical tools to come to standard data services. We will see increased sets of common services, both inside firms and cross-industry:

- Internally, the IT department will have to provide the *data operating system* of the company and secure the smooth integration at handover points to customers and service providers.
- Externally, we will increase master data services at the industry level. These will cater both to master data services and also to processes themselves, such as running security master data or data collection for client onboarding.

A sound information supply chain should allow for the following:

- providing easy access to all the information boundary conditions for transactions: master data on products, venues, and customers;
- ability to manage various frequencies of data;
- enterprise-wide consistent data standards—ideally aligned with standards at clients and suppliers;
- management of information handover points and securing data integrity as it flows between applications, or enters or leaves the firm;
- quality and completeness of data via validation rules;
- reformatting of data to feed internal/external applications;
- easy reporting and disclosure of data to downstream users;
- robust security and audit functions;
- scalability to meet enterprise requirements;
- definition and collection of summary statistics and KPIs and feedback loop to improve quality;
- data governance: ownership of data domains and change management procedures.

There are many reasons why it is difficult to create enterprise standard data services even when you have the attention of senior management. There are challenges in the following areas:

- Making the *business case* can be hard when many costs are hidden at departmental level. It is difficult to assess the impact of centralized solutions because of constant organizational reshuffles that make for difficult cost analysis and cost attribution. When the organizational structure is subject to change, users put a premium on local autonomy and will favor EUDA.
- *Stickiness* of existing infrastructure. An information infrastructure is normally incredible sticky. It is hard to change it because it is costly and because of the vested interests in the people who look after it (not

just vendors of content and software but definitely also the internal staff that has built up all the knowledge that will become worthless when it is gone).

- *Licensing* costs. Getting enterprise-wide rights to the content and the systems. Vendors of both software and content will aim to preserve additional commercial opportunities by putting in place limitations of use. Financial institutions that aim to achieve this face potentially very large upfront license costs.

- There are *political* challenges. There needs to be acceptance of a dedicated service team decoupled from individual business lines. The political situation in an institution needs to accommodate that. Organizations have managing directors for lines of businesses and everything is organized vertically. There are not often interesting career opportunities when looking horizontally cross-business. Only in case of external marketing of services do cross-horizontal jobs have revenue targets and commensurate bonuses.

- Users need to agree on *common terminology*. There will be different types of services in terms of extensiveness of coverage (which attributes), frequency (real time or batch data), and quality (manually validated, automatically checked, or just pass-through from the vendor). But even so, opinions may differ within one institution on asset type or industry sector classification.

- Users need to agree on *ownership* of content. When ownership of content passes to a dedicated central department, that department has to carry enough clout and be credible enough in terms of budget and staff to gain acceptance. When ownership is decentralized, it can fragment along product, matrix, or client segment dimensions. Unfortunately, these different lines of segmentation can intersect. Furthermore, politically it can be difficult to determine ownership of customer data as this is directly linked to a commercial relationship. More neutral information such as business holiday calendar information would be easier to centralize. Security master information would be somewhere in the middle.

- It is *time consuming* to get sufficient users and systems to tap into these data services even if you get acceptance of what they should be. Data integration solutions are inherently sticky: it takes a long time to get them into place but to get them out of place can take just as long. The organization needs to be willing to invest for a prolonged period of time and there need to be sufficient pickup of the data services downstream to justify the investment. Plus there is the issue of trust: users need to let go in some cases and hand control to a separate data function.

- *Technology challenges* of dealing with a heterogeneous downstream architecture with applications on different platforms. There are also often legacy systems that embed their own data model, which in some cases just does not fit with the envisaged common services.

In the next chapter we will discuss organizational aspects of data management—and what the conditions are on successful data management and quality improvement.

REFERENCES

EDM Council, 2015. Data Management Industry Benchmark Report. Available from: <www. edmcouncil.org>.

Gruber, T., 1995. Toward principles for the design of ontologies used for knowledge sharing. Int. J. Hum. Comput. Stud. 43 (5–6), 907–928.

King, B., 2010. Bank 2.0: How Customer Behavior and Technology Will Change the Future of Financial Services. Marshall Cavendish, Singapore, pp. 379ff–380ff.

McGilvray, D., 2008. Executing Data Quality Projects. Morgan Kaufmann, Amsterdam, p. 5.

Chapter 7

Data Management Organization

Chapter Outline

7.1 INTRODUCTION: CHANGING DEMANDS ON ORGANIZATIONS

Data is a nondepletable, nondegrading, durable asset. If managed correctly with good-quality checks and augmented via feedback loops of users it will be central to the success of any operation and its value will increase over time. To ensure this increase in value and to ensure this value is accessible to all stakeholders, firms have realized they need to change the way they work with data. They need to look at data in a new way, focus on best practices in owning and changing data, and put in place organizational models with the right mix of internal and externally sourced services.

A Primer in Financial Data Management. http://dx.doi.org/10.1016/B978-0-12-809776-2.00007-7
Copyright © 2017 Elsevier Ltd. All rights reserved.
225

In this chapter we address organizational responses and new sourcing models in the changing world of financial information. This will cover what organizations are doing to address some of the themes we discussed. These themes include:

- changes in how data is produced, shared, and sourced;
- changes in customer service expectations and reaction times when it comes to information provision;
- increased regulatory reporting requirements in terms of both strength of the processes and granularity of information;
- increased demands on cost-effectiveness.

Organizations have responded by reorganizing and often centralizing their data operations and information technology. They have created new functions such as data stewards and Chief Data Officers (CDOs). They are looking more critically at the amount of data they source. They are looking to make the client experience easier and more nimble. They are experimenting with new technologies to create efficiencies and improve information availability to users, regulators, and clients. Technologies used in these fields are often grouped under the labels "fintech" or "regtech." Regulators too take a keen interest to improve the quality of the data they get from the financial institutions and also to make their own processes more efficient. More effective ways to automate business logic can lead to massive cost gains—at the expense of jobs in risk, finance, control, IT, and operations.

In addition to using internal technologies, firms look to the sourcing model at large: what parts of the data sourcing and integration processes should be kept in-house and which should be outsourced. Firms are recognizing the value of information inventory and asset management: applying the same rigor in information management as in the management of other assets.

The data services landscape is rapidly developing with many firms investigating shared services for master data services on legal entities and financial products. The model is moving from simply outsourcing specific functions to dedicated service suppliers to the next level of shared services that blend technology and operations. The questions will be where the boundaries will be between federated and central internal functions and industry shared services.

In this chapter we start with a discussion on the field of data governance, that is to say, the ownership and change management on information. This is followed by an overview of different organizational approaches of data management. We discuss the traditional dividing lines of business (front office, risk, and operations) and IT and what the introduction of CDOs and data stewards could mean. This is followed by an overview of sourcing options for data management, both the traditional labor cost arbitrage model of outsourcing and also, more interestingly, industry shared services, utilities, and the reaction of traditional IT services and data suppliers. The main questions will be those of dividing lines and what the scope of data utilities could ultimately be, in other words, what functions will likely stay in-house and constitute core competence and differentiation. We conclude with an overview of change programs to make the transition to new sourcing or organization model and an outlook for the future.

7.2   INFORMATION GOVERNANCE

One of the key tasks of management is to look for the common good of the organizations and to transcend the here and now, and the specific. This is an especially thorny goal for data management. Reliable, accessible data is arguably the largest common good in organizations but there are significant forces of fragmentation at play.

Information governance looks in another way at the data management problem: not so much in terms of technology but by recognizing the social dynamics of it. In fact, it says technology is not the issue, if anything there probably was too much technology in financial services, or in any case, too many uncoordinated automation efforts. Data governance is complementary to and balances enterprise data management (Fig. 7.1) (drawn from Ladley, 2012).

A conceptually simple problem of acquiring, integrating, cleansing, storing, and publishing processed data has led to enormous cost and poor service to end users at best. Perhaps one of the major problems is that many firms consider information management a simple technology issue and do not invest the time, talent, or money, and have insufficiently considered the organizational preconditions for success.

Several organizational forces have caused the failure of an uncomfortably high percentage of IT projects (see, e.g., overviews of project fail rates at https://www.kpmg.com/NZ/en/IssuesAndInsights/ArticlesPublications/Documents/KPMG-Project-Management-Survey-2013.pdf). These include social complexity in a very large and diverse set of stakeholders involved in a project and

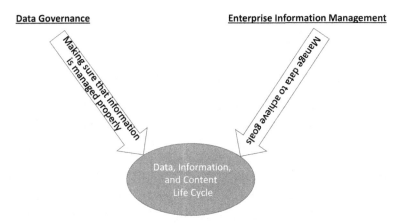

FIGURE 7.1   Data governance.

an overall "wickedness" of the problem (terminology in this sense introduced in Churchman, 1967). Wicked problems have complex interdependencies and hard-to-pin-down requirements. In particular, they have the following characteristics:

- You do not understand the problem until it is solved: what the problem is depends on which stakeholder you ask. This makes it difficult to define the issues to be resolved and to define success criteria. Different stakeholders will all believe you are solving a different problem: theirs. Each department will demand a data services group that supplies their personal data needs.

- The problem has no "stopping" rule. Given a constant flow of new reporting requirements, market evolution, and new information captured, there is no hard stopping rule. Criteria for success are on the ability rather than specific execution on reports or data types. In other words, data management is not about selling a specific project—it is advocating a new way of approaching the business and of processing changing data requirements.

- Solutions are not right or wrong and there is no magic wand, silver bullet, or easily repeatable recipe: solutions have to take the specifics of a business into account. Make sure you know what problem you are solving.

- The problem is unique and novel. The permutation of jurisdictions, product sets, client categories, competitive differentiators, and organizational cultures is unique for each company. Where are the similarities in the challenge and are they really the same? Whereas high-level technical architectures are all the same, implementations are all different. Rather than at technology, look at the business processes first and fix those if necessary. Streamline the business processes and you will already materially simplify the data needs (this "manufacturing"-type approach is already helping with the industrialization of the back office and part-industrialization of the midoffice, but by definition front office is where you make the money and here you need to celebrate nuances).

- You have only one shot at the problem. Large data transformation programs are significant investments and can be "once-in-a-career" opportunities to success or fail.

- There are no given alternatives. Cost pressures, and regulatory and customer requirements demand changes and cost reductions due to labor cost arbitrage have been made. Outsourcing in itself does not improve any process—it can also introduce additional overhead so there is no clearly identifiable third party that could solve the problem. In Section 7.4 we will discuss how this may change with emerging reference data utilities.

In an early important paper on the operational risk implications of poor data quality, Grody et al. (2006) identify a large number of discrete organizational structures within a diversified financial firm that could independently source reference data (see also https://www.capgemini.com/resource-file-access/

resource/pdf/Reference_Data_and_its_Role_in_Operational_Risk_Management.pdf). Independent data sourcing can be set up by business unit, geography, investment style, asset class, and so on.

Independent sourcing doesn't solve anything in itself though—it is just a reflection of separate budgets and short-circuiting central functions. A firm's different geographies, business units, and asset classes interdepend and rely on each other as much as on external data sets. They need different views but all rely on accessibility and confidence in data and there is still widespread dissatisfaction with finding and utilizing data [see https://tdwi.org/research/2016/08/improving-data-preparation-infographic/asset.aspx; dissatisfaction with finding and utilizing data is widespread: more than one-third (37%) of respondents indicated dissatisfaction with how easily they can find and utilize relevant data].

Cost pressures will cut through previous budget borders and will force departments to collaborate and align. This has created the new domain of data governance: processes to agree on data content and data change procedures. In addition, due to regulatory intervention, there is a larger awareness of the operational risks and the potential for fraud that poor data management creates.

Many larger financial firms put data governance procedures in place and have major data management improvement initiatives on their books. So what are they going to do differently now? In Fig. 7.2 we sketch the different activities mapped against the industrialization potential.

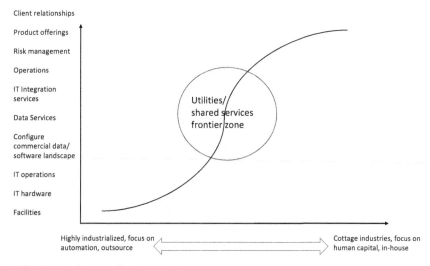

FIGURE 7.2 **Degree of industrialization.**

7.2.1 What Is Data Governance?

An important development the past decade has been a shift away from pure technological solutions. Firms recognize that tools are only as good as the organization that uses them and that data represents corporate memory and is an asset as well as a competitive differentiator in itself. The larger value of an information asset is found in the processes it supports and not in the information in isolation.

Money spent on large-scale processes, such as ERP and CRM is wasted if the data in it is not trusted. Firms know they have to wean off spreadsheets where possible, certainly where spreadsheets masquerade as production systems.

Following regulatory scrutiny and many costly IT project failures, firms have come to the conclusion that they have no choice but to implement more discipline with regard to data. They have seen that data can be highly politicized (who controls the data controls the firm) and realize they need a way out of that.

Data governance seeks to optimally facilitate cooperation on data-related matters across organizational silos. Data governance processes ensure all relevant stakeholders to represent business, operations, IT, risk, and finance are involved to share joint resources and that there is a common understanding to set the data agenda in terms of the following:

- Specify the process of sourcing, deciding on when and where to source, and how to quality proof and to distribute data from outside to the various end users.
- Set the agenda for changes on data sources, data models, business rules, and commonly used data.
- Prioritize changes to data sources and data management processes.
- Identify, resolve, and escalate issues that come up in internal data services.

In short, data governance can be defined as "making sure that information is managed properly" (Ladley, 2012, p. 10). The decision-making process can be supported by the data quality ROI methods discussed in Chapter 6. It is important that stakeholders of all the data sets in scope are represented ("stakeholders are those actively involved in the project or those whose interests may be positively or negatively influenced by execution or completion of the project and its deliverables," McGilvray, 2008, p. 69).

Data governance starts with a common organizational understanding of the principles of information management. These would look similar to Ladley's "Generally Accepted Information Principles" (p. 17) (Table 7.1).

Data governance means putting functionality in place *within* the existing organization hierarchy, not adding to it. To sustain good practices you need "organizational change management." By design, data governance is an enterprise program that is driven by the business use of the data, not by IT. It clarifies the roles, rules, and controls for the data assets. Successful data governance must melt into the fabric of the organization and *permanently* change the data culture.

TABLE 7.1 Generally Accepted Information Principles From Ladley

Principle	Description
Content as asset	Data and content of all types are assets with all the characteristics of any other asset. Therefore, they should be managed, secured, and accounted for as other material or financial assets
Real value	There is value in all data and content, based on their contribution to an organization's business/operational objectives, their intrinsic marketability, and/or their contribution to the organization's goodwill (balance sheet) valuation
Going concern	Data and content are not viewed as temporary means to achieve results (or merely as a business by-product), but are critical to successful, ongoing business operations and management
Risk	There is risk associated with data and content. This risk must be formally recognized, either as a liability or through incurring costs to manage and reduce the inherent risk
Due diligence	If a risk is known, it must be reported. If a risk is possible, it must be confirmed
Quality	The relevance, meaning accuracy, and life cycle of data and content can affect the financial status of an organization
Audit	The accuracy of data and content is subject to periodic audit by an independent body
Accountability	An organization must identify parties that are ultimately responsible for data and content assets
Liability	The risks in information mean there is a financial liability inherent in all data or content that is based on regulatory and ethical misuse or mismanagement

As Ladley puts it: "data governance is not a feature, it is the underpinning of all the possible solutions to use data better."

Perhaps ironically, the field of data governance has introduced its own technologies. These tend to apply the social media toolset and collaborative mindset toward improving data quality and getting agreement on definitions by sharing information as much as possible. Providers of specific data governance solutions build on these toolsets.

7.2.2 Data Governance Models

Historically, there have been clear organizational dividing lines that, while leading to ineffective data management, at least were clear. IT manages the systems and infrastructure and the business owns the data. Data management decisions can be very political so any data governance model needs to lead to decisions that stick if you want to come to increased quality and efficiency. Governance succeeds when the chain of responsibility and accountability is unbroken and when all expectations are documented and shared between all stakeholders.

There are different decision-making models in data governance. The Data Governance Institute (see http://www.datagovernance.com/; see also discussion about these models by Stockdale, 2014) distinguishes between the approaches shown in Table 7.2.

Whatever the governance model, the processes decide on the different steps in the data life cycle (Fig. 7.3).

The goal is the optimal use of data assets through a shared understanding and, where it makes sense, a single version of the truth and data integrity. You have data integrity if your data cannot be compromised. An Enterprise Data Governance program looks at the following:

- data integration
- data quality
- data model
- metadata

A good data governance program delivers (from Stockdale, 2014):

- clear policies and processes to manage data models;
- business data domain ownership defined with clear responsibilities for data in source systems, data warehouses, and data marts;

TABLE 7.2 Different Data Governance Approaches

Top-down governance and decision flows:	Bottom-up governance and decision flows:
Command and control	Some decisions made by individuals and
Clear links between executive-level data	via grassroots initiatives
governance and operational departments	Data naming standards driven by users
Alignment executive walk and talk; lead	Organizational ability to support grass-
by example	roots initiatives
	Stewards, data governance staff need to
	be visible and approachable
Silo-in governance and decision flows:	**Center-out governance and decision flows:**
Bring in representatives from multiple	CDO or CIO ask experts to specify data
groups to collectively agree on action	models and processes
Governance and stewardship councils	One or more centralized resources
Federated model can make sticky deci-	decide
sions but needs to be given authority to	Educate all stakeholders; consider
decide on models and processes	enough options
Representatives need to have authority in	
their own constituencies	

CDO, Chief Data Officer; *CIO*, Chief Information Officer.

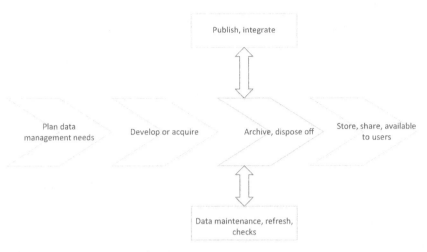

FIGURE 7.3   High-level data lifecycle management steps.

- establishment and implementation of an enterprise data dictionary to capture and implement data standards;
- policies and processes to support standardization of tooling;
- identification and assignment of sources for key data sets;
- appropriate KPIs to measure data quality;
- establishment of data quality mechanisms to measure and report quality throughout the life cycle of the data.

7.2.3   Data Governance and Cost Allocation

Because of the difficulty of quantifying the costs of data, an adequate assessment of the value of serviced solutions is complicated. It is difficult to judge a case for a BPO, outsourcing, an industry service, or indeed internal software solutions when the usage and overlap of data is unclear and when there is no agreement on data definitions. Data governance is therefore a prerequisite to undertake any large-scale change in data management.

The cost of the data supply can be charged to the consumers in various ways including:

- by the number of end users;
- by the number of primary sources used (directly correlated with content spend);
- by the number of consuming applications (correlated with the internal system integration work);
- by the number of files (correlated with data sourcing effort).

A proper governance process means that users that want to deviate from existing processes and sources will have to argue their case and be prepared to cover all the additional cost.

7.3 ORGANIZATIONAL APPROACHES

Requirements on data infrastructure and reporting increase while all firms seek a "best value for money" operating model that can serve the business cost-effectively. Many firms are looking at their organizations and their governance model, and have appointed new roles, such as a CDO and Data Stewards to establish firm-wide data governance.

One of the unique aspects of financial services is that, in financial services, the legal repercussions of poor data and poor data governance are enormous. What can make a difference in perspective is whether the focus of the EDM function is to *source information* (e.g., in the case of an asset manager or a bank) or to *publish information* (e.g., in a fund administrator, custodian, a data vendor). Should the delivery focus of the EDM function be to *internal* or to *external* clients? From an overall quality perspective it should not matter so much but there can be repercussions on what would be included in SLAs.

7.3.1 What Organizational Models?

As organizations seek improvements in data management, look to instill data governance, and create new positions, they also look at the overall organization. They have to answer questions, such as:

- What are the data priorities, and where are our biggest problems?
- What organizational models work best or are most appropriate to address them?
- What are the expectations of clients and regulators and what organization model works best?
- Which data governance model works best for our organization? What granularity do we need in decision making?
- How do we change to this new model that focuses on the longer term yet also unlocks all the data we have in house for common use? How do we avoid a proliferation of short-term initiatives with IT and operations left to stitch them together?
- What do we want to do in-house and what could be supplied by a third party? What are the scope boundaries of utility models?
- What is the impact the various stakeholders should have?
- Should we realign internal functions of IT and Operations?
- What should be the success criteria of data stewards and CDOs?
- Do we need a new central shared data services organization or should this be at business unit or country level?

Any discussion on new service providers or new tooling to be put into place can come only after all these questions have been answered *and* there is alignment on the answers.

Apart from introducing new data governance procedures, firms have already put in place models that aim to shorten the cycle of change and innovation. These include:

- combined technology and operations ("TOPS") models (these models see everything nonrevenue-producing as one service organization);
- combining IT development and operations/support in DevOps models to come to a new and faster software creation and deployment process.

Since no firm has the luxury to go completely greenfield, another approach is to innovate around the edges. This could be done by carving out specific areas of change that do not directly sit in the main production flow. Rather than dramatically overhauling the main infrastructure, new functionality is introduced in parallel in *sandbox* environments. A sandbox environment can be seen as a playground in which radical change can be introduced and tried out in small scale. Since the actions firms take to optimize the running cost of the existing infrastructure through clear change management, outsourcing operations, and shared data centers tend to put a brake on innovation, *sandboxing* is a necessity. New functionality coming out of the sandbox can be prototyped with small groups of customers and, if proven, elevated to the main environment and to all customers.

7.3.2 New Roles in Data Management: CDOs and Data Stewards

The recognition of the value of data management has reached senior management thanks to high-profile cases where a lack of data governance backfired and led to large fines or settlements, and thanks to cost pressures. On top of this, new technologies promise to get more out of the data to increase sales and improve customer service. They see that current needs remain unfulfilled due to backlogs, delays, and outright failures of IT projects and put in place reorganizations or appoint CDOs to spearhead improvements in data management.

Intuitively, the idea of data as an asset is appealing but is only slowly creeping into the minds of CFOs. It is difficult to put data assets on the balance sheet of a financial institution; yet the access to clients and knowledge of clients has the most value. More techniques to measure the value of assets are required. From a data curation process perspective, CFOs have the most experience in metadata management and data quality. Reported financial statements must be accurate and literally underwritten by the CFO. For the CFOs the data quality story may sound familiar but they may use different terminology.

The question is the following: what is new with the arrival of the CDO? Did we not already have Chief Information Officers (CIOs) and Chief Technology Officers? Is this merely semantics? What is different? CIOs run IT departments and CTOs often have a more architectural focus, so what was missing? The CIO title already contains the word "information." Why didn't the CIO focus on data? What different skills does a CDO bring to finish what the CIO supposedly started? There is more here than just the need for a fresh start and a new job title.

Data is the cross-section of the terms "information" and "technology" and the CDO (a good overview of responsibilities of different roles in data management is provided in Stockdale, 2014, p. 474ff) is much more a business function that:

- focuses on data governance, not on tools;
- looks at organizational change needed, rather than at IT operations or technology stack;
- understands data quality and can quantify its business impact;
- drives organizational change to improve the data management culture to come to a common understanding of data models, rules, and sources;
- advises on the best sourcing strategy using a combination of technology and service providers, internal departments, and industry utilities;
- brings data literacy and data management to the next generation of managers;
- is cross-functional and decouples data sourcing from data usage at the enterprise level;
- infuses the organization with best practices holistically on data management.

The other new role in data often introduced concurrently with a CDO is that of the *data steward*. Data stewards operate one level down from the CDO to help simplify data sourcing, streamline operations, and coordinate the data needs of different stakeholders.

Data stewards typically have responsibility for a data subject area across business processes and applications. Data stewardship has been defined as "an approach to data governance that formalizes accountability for managing information resources on behalf of others and for the best interests of the organization" (Ladley, 2012, p. 125). Data stewards look after certain data domains (e.g., client data segments, market data, legal entity information, product master data, risk data) on a day-to-day basis, inventory the information needs of users, create and maintain a data model, and define, agree, and look after quality rules. They also keep in frequent contact with stakeholders, communicate on changes, and set up data policies. Data stewards form a data stewardship council. A potential organizational model for federated data governance looks like as shown in Fig. 7.4 (based on Ladley, 2012, p. 134).

Data stewards can operate at tactical or strategic level and their job can blend the roles of (after Ladley, 2012, p. 126):

- Information executive. This includes the approval of information principles and policies. Monitoring data scorecards. Casting vote in data governance decisions on their domain.
- Information manager. This includes a deep understanding of all use cases pertaining to the specific data domain and decides who can use the information. Ensures correct day-to-day management of data assets.
- Information custodian. This includes quality audits and policy compliance. Direct execution of work in policies and procedures. Education of data specifics to the business.

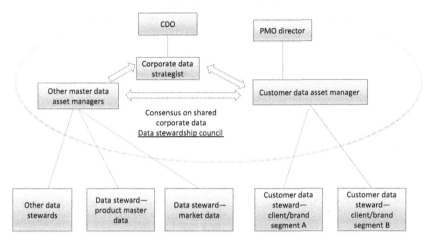

FIGURE 7.4 **Federated data governance.** *CDO*, Chief Data Officer.

The CDO role is fairly new in most companies and the specific to-do list of CDOs will differ by firm (Gartner has predicted that by 2019, 90% of large organizations will have hired a CDO, but that of these, only 50% will be seen as successful; http://www.gartner.com/webinar/3283528). The success of the role depends on organizational preconditions. Although data management has been with IT efforts since the beginning, it has tended to play second fiddle to application delivery. The success of a CDO or any data management initiative starts with the recognition of data management and data governance as a necessary discipline. CDOs need to be able to focus on data management best practices and organizational rather than IT change, and have to report to the business.

Too much can be expected of a CDO if the role comes with very few staff and no direct influence over IT and operations. The risks are then that the role does not get beyond good intentions and wishful thinking. Given how central access to reliable data is to everyone's job, the CDO's own job spans an entire organization and every level of project from tactical to strategic and every stage of change.

A CDO has to wear different hats and juggle different perspectives (Fig. 7.5) including:

- A risk view: what could we be missing? What data could be lost? Where could we be in regulatory trouble? A need to protect the corporate reputation, manage uncertainty, and clarify the areas of uncertainty. Informed risk management allows making conscious choices about uncertainty.
- Operational view: how do we source and treat data only once and get the data in correctly the first time? Also, how do we prevent bad data from entering the organization's IT stack and polluting other applications leading to reduced trust in the data?

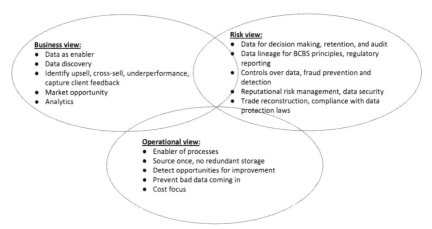

Business view:
- Data as enabler
- Data discovery
- Identify upsell, cross-sell, underperformance, capture client feedback
- Market opportunity
- Analytics

Risk view:
- Data for decision making, retention, and audit
- Data lineage for BCBS principles, regulatory reporting
- Controls over data, fraud prevention and detection
- Reputational risk management, data security
- Trade reconstruction, compliance with data protection laws

Operational view:
- Enabler of processes
- Source once, no redundant storage
- Detect opportunities for improvement
- Prevent bad data coming in
- Cost focus

FIGURE 7.5 **Different CDO perspectives.** *CDO*, Chief Data Officer.

- Business view: how do we optimally use the data to serve our clients, and get them the products and services they need?

Clear expectations about what you expect to change are critical when bringing in a CDO. A very good taxonomy of different styles of CDO has been created by Jennifer Ippoliti:

- **Regulatory.** The focus is data management catering to regulatory compliance and risk. This type of CDO could report to the Chief Risk Officer.
- **Analytics.** The focus is on new ways to crunch data and new problems to solve. This type of CDO could report to the CIO.
- **Data Quality/Data Governance and Policy.** The focus here is on addressing systemic data quality problems, and instilling changes in procedures and/or organizational models. These types of CDOs typically report to the CEO, COO, or CIO. This is also the most widely held perception of what a CDO should actually do: improve data quality, and access and data management procedures across the firm for all stakeholders. When reporting to the COO, there could also be a strong efficiency component to it—reduction of cost in direct data spend or headcount by centralizing data flows and streamlining data quality procedures.
- **Revenue.** This CDO category would report to the Chief Marketing Officer and has a specific focus of looking at what new revenue streams can be created with the data. This kind of CDO would also look at how a firm's data could be monetized directly (e.g., by creating new data products to license either directly to users or to data aggregators).

Unsurprisingly, the best CDOs are those who can blend different flavors depending on the circumstance. The field of CDOs has quickly developed during the past 10 years as executive management has gotten a clearer grasp of the seriousness of the industry's data issues and needs. Simultaneously, data quality metrics and total data cost metrics are evolving and help prove that CDO-instigated initiatives create specific, measurable, and forecastable business benefits (from http://www.forbes.com/sites/sungardas/2014/11/11/the-5-flavors-of-chief-data-officers-cdos/#cdd112481e07).

7.3.3 Turning Employees into Responsible "Data Citizens"

Putting in place data governance spearheaded by a CDO and operated by data stewards is only the start. The success of such undertakings hinges on convincing all stakeholders (and external service providers) to cooperate. The job of the CDO includes education and evangelizing and to raise the overall data literacy of a firm. Literacy includes understanding Ladley's Generally Accepted Information Principles but also means you understand data models, concepts such as audit and lineage, the different maturity levels of data management, and their impact on an organization from a cost, a revenue, and a risk perspective. Widespread data literacy means firms can better judge IT management, business processes, and IT architecture and can come to better sourcing decisions.

A CDO will have to convince staff to behave like responsible data citizens by seeing the benefits of data as a common good, by avoiding local data sourcing and data management solutions, by sharing information, and by contributing data and insights to the pool. Incentivizing people to build for the mid or long term is part of this. Hiring policies, educational programs, policies and procedures, sharing results, engagement, and collaboration tools to harness data knowledge in the company and above all communication are all part of raising the data literacy and turning an entire company into a highly data-aware organization that can evolve with changing business, client, and regulatory demands.

7.4 OUTSOURCING AND SERVICE OPTIONS IN DATA MANAGEMENT

Firms that have well-functioning data governance procedures in place and that know where their costs and pain points are can look at changes in the sourcing model. Fortunately, the options available to source information and information management sources have become increasingly diverse.

The manufacturing industry is often held up as an example of an industry that, unlike financial information, has gotten its supply chain and cost management right. All parts are standardized and suppliers are closely integrated into the manufacturing processes. Lean operating models have brought operating efficiencies and Six Sigma has brought close attention to quality at each stage.

Financial services firms look for manufacturing-style efficiencies in their processes. The flip side of this is that they will have to change the way they work, accept standardization of data models, and redraw the boundaries of what happens inside or outside the firm and thereby surrender part of the control over the end-to-end operation. There have been different reports on cost pressures in the banking industry and the need for standardization (https://www.rolandberger.com/publications/publication_pdf/roland_berger_state_of_european_banking_industry_20140721.pdf). Reports have argued for simplification in terms of operational fit, organizational streamlining, and refocusing the footprint (including carve-outs/outsourcing, consolidation, and shared services).

Many firms seem to have already started on this path of simplification. Data is increasingly bought and managed at group level and IT and operations are brought closer together. The next step is to bring in third parties—either dedicated 1:1 providers or providers of shared services (1:*N* models). In this section we discuss areas that could be shared and what organizational models are most appropriate.

Engaging with third-party suppliers forces you to have your own house in order. You need to have a deep understanding of cost and pain points to be able to successfully negotiate a service contract with a supplier (Fig. 7.6).

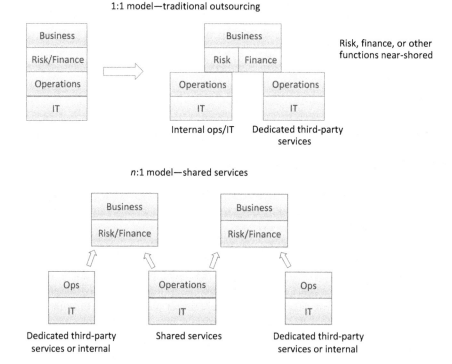

FIGURE 7.6   **Outsourcing and shared service models.**

7.4.1 Different Service Models

Common problems lead to new products or services. They have led to packaged software applications sold under license or common services such as audit and valuation. The new focus on standardization of processes has led to many new packaged services in IT and data operations.

Simply put, external service providers can either take over part of your existing operation (we call these 1:1 providers) or can already offer a service to others that you subscribe to (we call these 1:N providers). There are also intermediate models where, for instance, only hardware is shared but dedicated instances of software are operated. Different approaches will come with different KPIs and commercial models. Common service models (Table 7.3) include:

- In an *Application Service Provider* ("ASP") or *Managed Service*, software is run by a third party—often the supplier of the software. This can imply that proprietary content leaves the building. If a separate instance of the software is hosted for every client, customization of the service is still possible.
- These solutions are often based on a vendor's enterprise software offering and provide a dedicated instance of the software for each client, sometimes in the cloud, as well as the ability to configure the software to meet business needs. Expected benefits of hosted solutions include reduced total cost of ownership, increased speed to implementation, enhanced data quality, and a lower requirement for technical skills. The burden of deploying and maintaining software in house is removed.
- *Software as a Service* ("SaaS") refers to software services that are designed from the outset to be delivered over the internet. An individual application is hosted centrally. Software is not customized and the client manages its own content. This can work for end-user applications but is rare as part of a core business process in financial institutions. SaaS models can lower the hurdle for institutions as costs can move from a CapEx budget to an operating expenses model. They can also lead to a change in departmental charge-backs and different attribution of costs. The SaaS model of renting assets that were previously under CapEx budget has spread over infrastructure ("IaaS") and platforms ("PaaS") leading to the phrase "Everything as a Service."
- *IT Outsourcing ("ITO")*. The operation of a firm's technology infrastructure is handled by a third party. Benefit could be that the service is run out of a lower cost base and that data centers are shared.
- *Business Process Outsourcing ("BPO")*. Operations such as order fulfillment or customer service are handled by a third party. There is benefit because of lower labor cost base and potential technology sharing.
- *Industry Shared Services*. In this case specific business processes are serviced by a common service that every firm can subscribe to. The financial services industry has long had utilities for price discovery, trading, and clearing and settlement (clearinghouses, exchanges, and Central Securities Depositories) and specific functions in posttrade processes have given rise to

TABLE 7.3 Summary of Service Models

	Scale	Hardware	Software	Operations	Pros	Cons
ASP/managed service	Application	Shared	Dedicated	IT operations dedicated or shared	Benefit from shared infrastructure, business continuity advantage if firm is small	No option to customize, change requests can be costly or take a long time
SaaS	Application	Shared	Shared	Shared	Rental model	No option to customize
ITO	Firm	Can be shared	Dedicated	Dedicated (IT operations only)	Lower cost	Overhead, switching cost
BPO	Firmwide	Included in BPO contract	Dedicated	Dedicated (IT and business operations)	Lower cost	Overhead, switching cost
Industry shared services/utilities	Specific business processes	Shared	Shared	Shared	Lower cost, standardization of info exchange with regulators and peers	

ASP, Application Service Provider; SaaS, Software as a Service.

custodian banks, fund administrators, and agents for securities lending and collateral management. What is new is a focus on services for *operational data support* for processes that to date have stayed in house. Utilities have sprung up, for instance, to support KYC processes and reference data management (see KYC utilities from Genpact/Markit and Clarient from DTCC; in reference data Smartstream has launched the Reference Data Utility with Morgan Stanley, Goldman Sachs, and J.P. Morgan). The difference between *utilities* and shared services is largely one of expectations on the commercial model. Utilities would be industry owned and governed and the commercial model would be conservative and could be cost recovery or cost plus.

7.4.2   1:1 Models

In models where you take a dedicated "lift and shift" service, there will be a bilateral contract. The impetus here is often to benefit from a supplier's lower cost base. Service providers operate out of lower labor cost countries and their staff may not have the same amount of benefits that financial services staff used to have. Scale of IT or Business Process Outsourcing can be department, division, or entire firm level. Larger firms tend to have two or three preferred suppliers.

Outsourcing has been going on since the late 1990s after the common adoption of the internet. Meanwhile, larger firms have captive organizations in the same jurisdictions as service providers (India, East Asia, Eastern Europe) for application development, maintenance, and operations and the potential for further cost reduction is limited.

A prerequisite on successful outsourcing is knowing your cost base and baseline. The firm needs to have a common expectation on what metrics to capture in an SLA and what are achievable targets, to avoid a "your mess for less" broken model locked in concrete by an SLA in the hands of a third party. With increased automation and the rise of machine learning, the percentage of staff cost as a portion of total cost goes down. This will reduce the need for labor cost optimization.

7.4.3   1:N models; Industry Shared Services and Utility Models

Data utilities use a shared platform that handles data once and disseminates it many times. Given that cost benefits from ITO and BPO have been realized, the potential for efficiency is looking for the largest noncompetitive common denominator in firms. The huge redundancy in the industry is of processes repeated in every company that do not lead to competitive differentiation. Processes that can be mutualized

- have to be scalable;
- have to be repeatable;
- have to support business processes that are very common and ideally the same across jurisdictions;

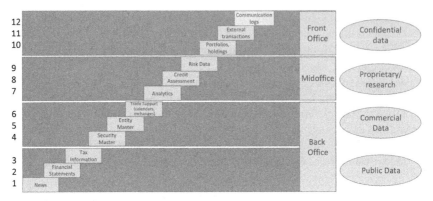

FIGURE 7.7 Data types ranking in terms of internal value-add and proprietary nature.

- should not rely on confidential information;
- should not be a competitive differentiator;
- should not directly report to the regulator.

Just as the advent of the internet took down geographical distance and enabled ITO and BPO, improvements in workflow and data integration technology help shared services.

The question is, looking at the data types discussed in Chapter 2, and the business processes reviewed in Chapter 3, the following: which data types lend themselves to be mutualized? What combinations of data curation and business processes can become packages that are sufficiently common and standard to be bundled as utility services (Fig. 7.7)?

Data utilities have focused on back and middle offices. Examples include:

- valuation services (sourcing and validation of price data and valuation software is shared);
- corporate action services (aggregation of corporate actions data, cross-referencing, and reporting is shared);
- security master data cross-referencing, cleansing, and the delivery of a bundled masterfeed;
- collection of entity information to support KYC and client onboarding.

Some data utilities constitute an intermediate step between aggregators and the user firms, acting as "super" aggregators creating one format and bundling content with data validation services. This standardization and validation of aggregated data may eliminate a lot of the work in the internal part of the supply chain. It could be offered as a managed service as arguably many organizations do the same thing in-house, possibly bundled with other Business Process Outsourcing services such as cleansing of data according to client-defined business rules. These cleansing rules could vary from client to client making the scalability of these kinds of offerings somewhat difficult as we move closer to the area where the core business of financial institution starts: the value-added proprietary business rules and models. Example client-dependent rules include:

- Which data vendor to use? Not every client will take the same sources and different clients may have different perspectives on the relative quality merits of vendors for various market segments.
- Which market makers to use to get snapshot prices?
- Which cleansing rules to apply to flag outliers and to fill gaps in time series?

The more valuable utility services will cater to client use cases and will solve "last mile" integration problems, for instance, by integrating with commonly used software applications or by supplying filled-in templates with all required regulatory information.

Unlike 1:1 suppliers, utility providers will work with standard contracts and standard service levels. This could mean that the internal processes need to change to benefit from a shared model. The larger the scope of the service, the larger the governance demands would be. Firms will always have to keep a backup plan in mind. Some service providers will be industry owned which would make concerns about lock-in smaller. Furthermore, some large service providers may become heavily regulated themselves if their role is of systemic importance.

7.4.4 Redrawing the Borders of the Back Office: What's In and What's Out

More and more of the back office will consist of shared services as consolidation of the industry continues. Firms will have to ask themselves the following:

- What types of suppliers are best suited and reliable for shared services?
- What are the criteria to evaluate them?
- What is the right construct for a supplier relationship? What outcome-based metrics make sense for the specific service? These could contain the KPIs and data quality metrics discussed in Chapter 5.
- What commercial models make sense? These can be a combination of fixed cost plus a variable part based on shifts in demand or deviations from expected service levels.

The potential for outsourcing declines with the customer proximity of the function: cost centers can be outsourced but you won't touch the secret sauce of the business (Fig. 7.8).

The boundaries of shared services will be defined by constraints in:

- Confidentiality. Customer data will remain in-house for the most part bound by data protection laws on personal data. However, a substantial part of basic data supply on company information such as address and legal structure can be outsourced.

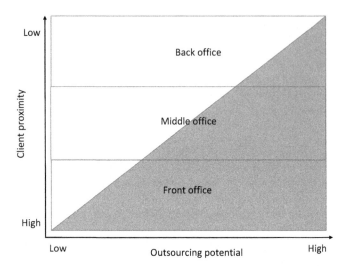

FIGURE 7.8 FO/MO/BO industrialization.

- Commercial sensitivity. Any information on credit assessment and valuations underlying products the firm sells has to be kept in-house.
- Regulatory obligations. Some regulators demand certain data is kept in house. Also, the end point of regulatory reporting has to come from the financial institution itself.
- Dependencies on third parties. Some third-party service providers on whose services you rely may wish to deal only directly with you—thereby requiring you to integrate their services yourself.

7.4.5 Implications for Service Providers

Incumbent suppliers of software, data, and services need to adapt to the world of "everything as a service" and the industry move to shared services. It means they may have to either provide shared services themselves or be willing to provide their services to third parties instead of to end users (Fig. 7.9).

1. Consumers demand faster response times and interact often with providers on a per transaction basis.
2. Financial services firms have to be more responsive to get and keep customers. They are faced with a much more diverse sourcing landscape and fintech firms provide opportunities as well as threats.
3. New intermediary fintech firms will get in between banks and customers and can become a parallel channel.
4. Traditional data and software firms not only supply to their traditional financial services customers but can also find new customers in service providers.
5. New XaaS providers decompose the problem space, can specialize in certain use cases, and often sit in between consumer-facing firms and traditional data and software suppliers.

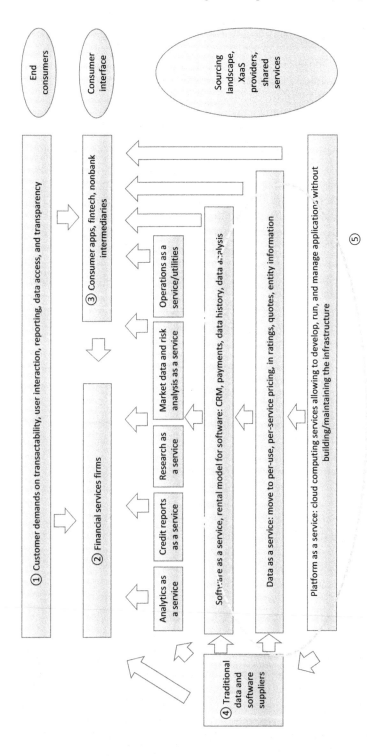

FIGURE 7.9 Sourcing disruption, to an XaaS sourcing landscape.

Different types of companies that could fulfill a role in managed data services include enterprise data providers, existing market infrastructure companies such as clearinghouses, exchanges and depositories, and large custodians, large fintech conglomerates, and large IT services companies that have a significant footprint currently in firms.

Specifically for content products, the advent of utilities will reinforce the following requirements:

- Products need to be set up and delivered with the *anticipated workflows* of the content in mind. How users across the institution consume the information needs to be accounted for in the product. Content providers have to anticipate the integration and data enrichment work that still typically happens within a financial institution.
- Products need to include quality assessment, confidence scores and lineage information as well: they should be clearer and more explicit *about what happened up to that point*. This can include information on the confidence the vendor has in the accuracy and completeness plus also timestamps on when the content item was originally picked up.
- Products should become more open and flexible with regard to the ability to take in information and conform to data standards in identification, classification, and formatting.

In general, data products need to cater to downstream use cases and contain appropriate metadata for the downstream user to judge whether and where they will use it. This could include:

- *combining data*, not just offering bid/ask prices from many venues but also offering consolidated prices as per a client's instructions leading to mass-customized products;
- *tailoring delivery*, for example, taking request lists based on customer lists and securities of interest and delivering data in a format that can be directly picked up by the customer applications;
- *deriving data*, not just offering prices but also directly offering the piece of information the customer needs to value its books and measure its risk;
- *APIs and on-demand data*, proactively sending out notices when something changes, for example, the rating agencies and news services that can give alerts on predefined business critical events.

7.4.6 Next Developments; N:M Models?

We discussed models to get efficiencies stemming from suppliers operating in lower cost bases and, more importantly, coming from different degrees of sharing from infrastructure to operations. Another development is that financial services firms have also recognized the value of the data that they create as a by-product of their main business or that they have to report externally due to new regulation. Especially market infrastructure companies such as exchanges and more recently custodians and depositories have set up data commercialization

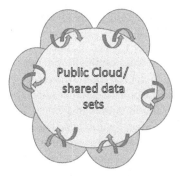

FIGURE 7.10 Public/private cloud: sunflower model with petals as private data areas around public data set.

units. But banks and asset managers too may create valuable content that they can put to use outside their own firm.

In cases where there is not an ultimate source such as a prospectus, firms could pool data together and collectively create a data service for the industry. This can include credit scoring, valuation, and quotation. This could lead to an $N-M$ model where data solutions are effectively crowdsourced.

Data pooling services have existed for a long time: enterprise data providers and exchangers collect quotes for financial products, bundle this, and sell this as a feed. More recently, firms have had to disclose traded prices in OTC products to trade repositories. Although data still needs to be integrated, this task has become easier and cheaper with more powerful ETL tools. One effect of this is that data originators will get more bargaining power and control over the downstream usage of their data.

In Fig. 7.10 we show the interplay between public and private clouds where firms submit information to and consume from a shared infrastructure.

Especially in regulatory data such as granular credit information and risk factors, banks and asset managers could benefit from these pooling models that could either be run by current enterprise data providers or be set up in parallel. Firms would tap into the public cloud and submit their corrections to public master data or contributions to new public master data. Public cloud data sets include product master, trade support data, and entity master. Firms keep supplementary, more sensitive information in their private environments including credit information, business relationship information, and risk data.

7.5 CHANGE MANAGEMENT PROGRAMS FOR SHARED DATA SERVICES

Change management is any type of activity coordinated to facilitate the implementation of a project and help ensure its success in the organization. It has to take place for all degrees of change, from implementing a new system, internal reorganization, or redrawing the borders of the supplier landscape. The job of the CDO and of data governance is to support the reuse and sharing of information and to come to a common understanding of terms. In other words, to organize and optimize the business data architecture discussed in Chapter 3.

Data management's task is to feed business users and applications with timely, quality data to make them optimally productive. To get there, the data management function has to get the right technology, projects, culture, budget, and metrics in place. Budget processes favor silos and business case–specific projects. However, *especially* in data management you need a budget process that contains incentives to reuse data assets. The effectiveness of automation is negatively correlated with the number of different budgets and independent decision-making points when it comes to starting IT projects or buying data, in other words, the willingness to share infrastructure. In budget sessions, the prioritization of projects and ordering of them should be influenced by the *potential to generate data reuse*. Projects should be ordered such that each project needs to add only an incremental amount of data to the existing data set. This will reduce the investment needed for each new project as they progressively fill in any remaining gaps in data (from Brose et al., 2014).

Risk and finance departments have to cross silos, since they are the funnels where many data flows come together before being published out to regulators and the investment public. The normal contention for scarce resources and cross-project coordination issues tend to be much larger for risk and finance compared to for business-silo–specific projects.

Many firms look at large-scale modernization and improvement, driven by a need to streamline data management for internal cost reduction and improved customer service, and also simply to meet current and future regulatory reporting requirements. In larger-scale change management we distinguish between

- transition: the handover of the data management function to a new (service) provider;
- transformation: the rationalization and consolidation of a proliferated legacy IT master data management stack.

7.5.1 Best Practices

Before kicking off a project, it is critical to define success criteria and a clear roadmap. To engage users that will benefit from the project, business impact techniques discussed in Chapter 6 can be used. Often anecdotes rather than just numbers can be of powerful help to remind everyone of why the project is there. There can be a lot of emotion in change and users should not be treated as spreadsheets. Understanding what drives or hinders end users is a key requirement. Project managers have to look at a day in the life of the direct data users to answer the "what's in it for me?" question when selling the project.

Any data quality improvement process has a number of preconditions to be successful that include:

- accountability
- root-cause analysis

- prevention
- ongoing monitoring
- continuous improvement
- communication

In any change project it is important to keep a risk register and to log and communicate all decisions. Overall program risk assessment for the different projects has to be tracked along different perspectives, including people, process, systems, regulatory, and organization.

Locking down requirements for data management can be especially hard for financial and risk analytic transformation projects given the pace of regulatory evolution. Also these projects cut across business lines and typically have a horizon in line with regulatory change, that is, 2–4 years. Scope creep is very common but is one of the biggest causes of project failure. On top of that, the decades-old legacy IT infrastructure means that an understanding of the current situation is often a nontrivial task.

The key to change management, as in data governance, is stakeholder management. Stakeholders in a change program can include business users, technology development, technology support, operations, vendor management, procurement, and project managers. Social complexity and organization structure has to be overcome by a clear change organization, project methodology, and frequent communication. Users that are left out and not consulted can easily frustrate a project.

There are many different project management methodologies designed to control change through effective measurement of progress, communication and alignment between stakeholders, risk management, definition of responsibilities, and creating the best possible team. A first rule of thumb is to agree on the methodology.

A change organization needs proper governance that would include regular steering committee meetings and, as required, one-to-one meetings with individual stakeholders by the project management. The key to change is very regular communication to show progress and sustain buy-in and any project plan needs to include a communication plan. Communication can take many forms on top of formal meetings including town hall meetings, newsletters, and stand-ups. Progress has to be tracked formally too and a common understanding (and timely communication) of achieved milestones against project budgets is critical. Financial KPIs can include budget to actual spend, estimates to completion, and deviations against original plan as well as earned value analysis (EVA) (this looks at actual useful work completed rather than money spent; see Brose et al., 2014, p. 400).

Stakeholder management includes answering the "what's in it for me?" question. Knowing how *data literate* stakeholders are as well as their organizational responsibilities, governance, budgets, and RACI matrix are all basic input to stakeholder management. Good stakeholder management includes a *stakeholder mapping* exercise that assesses the relative influence and commitment of each of the stakeholders. In other words, raising the awareness of the politics to get the required support and to spot risks early. This will help the management team understand where they need to spend their time in terms of managing organizational change (see Brose et al., 2014, p. 393 for a good discussion on stakeholder management and p. 417 for an overview of project implementation steps).

Specific roles in change management include a Program Manager who keeps the oversight and tracks dependencies across different projects and Project Managers who focus on individual projects. We provide example roles and responsibilities in Section 7.5.3.

Before embarking on any change, whether a complete reorganization of data services or a platform implementation, the target model needs to be clear. For this, the current architecture but especially the future information architecture needs to be understood. Information architecture "is a detailed description of data flows between the various systems that will be required to deliver the requisite information to achieve the program's requirements for calculation and reporting" (Brose et al., 2014, p. 408). A data gap analysis assesses the firm's readiness in collecting data required for calculations and the reporting required by the program.

A data gap analysis with regards to the business architecture sketched in Chapter 3 is required as well as a complete *inventory* of required data logistics. With this we mean an overview of what data needs to be where and at what moment and at what frequency in time, together with any specific requirements or boundary conditions (e.g., latency and accessibility). This will lead to an operational data needs calendar such as the example below(Table 7.4).

The data needs need to be clear as a precondition to get users what they need and to look for overlapping requirements to cost-effectively source. Also, a data inventory can help divide the data in different domains (market non–real time, real-time, product master, client master, transactions) that correspond to the remits of data stewards.

Lack of a complete understanding of the last mile integration is typically the source of delays or even project failure. Vendors make a living packaging up specific use cases but the value of proper change management is to embed these packages into an existing infrastructure. To ensure services meet the requirements,testing includes the verification of the proper transformation of the data, the integrity and rejection of poor data, the coverage and data access.

7.5.2 Phases of Change

Any project that has different phases and project management methodologies can differ in the granularity and naming of the phases. Typically, phases of a program include (see Brose et al., 2014, p. 406, Table 25.1, for more information):

- high-level business needs analysis
- detailed analysis

TABLE 7.4 Example of Operational Data Needs Calendar

Data set user	Frequency/time	Contents	Source/distribution	Quality constraints	Permissions
Market risk	Business days at 5 p.m.	10-Year historical data (close prices) for expected shortfall	Bulk file	Sources cannot include the bank itself	Public
Operations	Business days at 4 p.m.	Security master delta file: all changes in securities in which there is a client position	JMS queue	Data sources from Bloomberg, Thomson Reuters	Within operations
Collateral management	Every business day at 11 a.m., 2 p.m., and 4 p.m.	Interest rate curves for EUR, USD, DKK, CHF, SEK, JPY, and GBP	Bulk file	Check curve consistency, use three sources for each curve point, and take average	For distribution to counterparties
Corporate actions	Every business day at 11 a.m., 2 p.m., and 4 p.m.	Overview of all corporate actions of all securities in "of interest" list	JMS queue, separate queue for choice events. Format using ISO 20022	Flag "choice" events for which a customer response is required	
EOQ market data	Close prices to close the books at quarter end		JMS queue	Special rules and manual check	

- design
- implementation
- operationalization
- continuous improvement

In the subsequent text we provide examples of steps and responsibilities in a change program involving a major change in operations where a financial institution outsources a material part of its data operations to a third party, for instance, an industry shared service. This necessarily goes beyond attention areas in internally focused projects such as reorganizing IT and operations or implementing vendor or in-house–developed applications. For instance, in material outsourcing projects additional risk management, business continuity, and HR aspects come in.

Before embarking on such a change program, due diligence and analysis of the "as is" current state will be undertaken to come to detailed planning, assumption validation, concluding commercials, interim and target operating, and architecture model definition (Fig. 7.11).

This due diligence, scoping, and assessment phase of the project provides essential information to estimate the effort, risks, schedule, and sequence of activity in the phases that follow. Also, the Program Management Office ("PMO") and Project Management staff will be identified. Due Diligence can include different work streams such as:

1. Operational Stream
   a. inventory of current data operations and cleansing/mastering/distribution systems;
   b. assessment of current business continuity measures in place and gap analysis against what is required;
   c. description of the capacity plan for active and ongoing projects;
   d. inventory of technology stacks, incoming data feeds, reporting, and downstream consumption requirements;

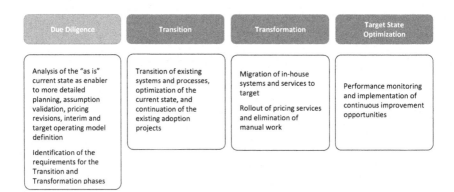

FIGURE 7.11 Change management steps.

    e. details of business rules in operation;

    f. information gathered including sources, reports, spreadsheets, downstream systems, delivery methods.

2. Personnel Stream

    a. capacity plan for taking on attrition;

    b. Staff Retention Plan and Staff Redeployment Plan;

    c. transfer of staff;

    d. documentation of roles, responsibilities, and organization chart.

3. Functional Stream

    a. analysis of inbound data, coverage, outbound data, data logic, downstream dependencies, continuity assessment;

    b. Gap Analysis of business rules across all data categories;

    c. optimization opportunities, documenting the overlap between interfaces, and business rules in legacy systems operated by the Client.

4. Data Supplier Management Stream

    a. review of data supplier contracts;

    b. establishing third-party processing agreements;

    c. optimization of data sourcing—cost analysis of aggregator/distributor contracts compared to direct to source arrangements;

    d. management of new data requirements.

5. Client Integration stream

    a. impact analysis of inbound/outbound interfaces, technology, and infrastructure;

    b. planning to cope with additional requirements;

    c. continuity assessment.

6. Platform Stream

    a. systems inventory;

    b. migration strategy;

    c. legacy systems support and maintenance arrangements;

    d. delivery of derived pricing requirements.

For each of these streams, an assessment is made that can include a series of high-level workshops. These serve to provide the client and the company teams with sufficient information to determine and analyze the status of operations in the client and begin the execution planning for the Take-On and Transition phases.

Assessment workshops comprise one or more of the following elements:

1. End-to End walk-through of "as is" current state operations

    a. staffing, key staff and related retention plans, succession planning, personnel issues, premises, supporting technology, planned changes;

    b. organizational charts, roles, responsibilities;

    c. processes and procedures—documented, undocumented, institutionalized, or with key staff;

    d. costs.

2. Set-up Information
   a. Systems and operations supporting legal entity data, parties, counterparties, instrument, etc.; organizational systems for budgets, reporting, personnel data.
3. Process area workshops
   a. For each "as is" operational area identify and document standard rules, exception rules, changes, reversals, data issues, and risks.
4. Facilities
   a. Premises, inventory of contents, office equipment, compliance with workplace regulations, insurance, issues, changes already underway, sublease or rental arrangements, workstations.
5. Personnel
   a. Organization charts, HR information, contractual obligations or conditions on change of ownership, alignment of packages, benefits and procedures, personnel issues.
6. Legal
   a. Known, active, or suspected legal actions or undertakings.
7. Costs and expenses
   a. Travel, benefits, pay reviews, budgets for changes, projects or systems, premises changes.

Using information uncovered in these workshops, this next phase establishes the best sequence of actions. For example, it may be best to implement according to a specific geographical plan or by specific business units, taking into account urgent issues and ease of implementation, or, where appropriate, by identifying quick wins.

These planning phases result in an agreement on the sequence of implementation steps and the approach to be used. Milestones are established with dependencies between different activities and deliverables identified and planned for. Resources from the client, the service company, and potentially a third party to assist are determined. A retention plan and a redeployment plan are established with the company HR department.

It is likely the project will be of a duration and scope such that there will be a set of plans to break the project down into achievable deliverables; these plans are then incorporated into an overall program that states all assumptions, risks, gaps and dependencies. Plans include sufficient time to "run up" staffing, premises, and equipment or tools to support its execution. In Table 7.5 is a set of typical transition deliverables.

7.5.3   Governance

The PMO involves support, reporting, and control objectives and should be a proactive program management structure ensuring that the program gets delivered on scope, on time, and on budget with the required level of quality. The PMO ensures an overall coordination and manages cross-dependencies. In case

TABLE 7.5 Change Management Deliverables

Deliverable	Description	Date	Acceptance criteria
Governance	Implement governance model Create program organization Establish PMO function (personnel and processes)		Terms of reference for all governance bodies signed off by client and service provider PMO handbook signed off by client and service provider All PMO resources mobilized and in place
Commercial agreements for remainder of program and target state	Formal contract and associated schedules		Signed master agreement Service levels agreed and signed off Signed-off processes for agreeing changes to the above
Documentation	Complete technical and operational documentation Dependency on client subject matter expert ("SME") resource availability Gap analysis		Workbooks created for all processes, systems, and data feeds to mutually agreed completeness Signed off method for validating that service provider data feeds match existing data feedsSigned off "as is" and "to be" gap analysis complete with proposed solutions
Plan and strategy for transition and transformation stages	Detailed plan for transition phase and in progress plan for transformation phase		Strategy for systems and operations in Transition and Transformation stages Program plan with mutually agreed detailed dates, key milestones and deliverables, client and service provider resource requirements Risk profile and mitigation plan, Transition and Transformation financials
Exit and rollback plan	Plan for termination of program or exit from BAU service after completion of program		Exit and rollback plan signed off by client and service provider

issues are not resolved in due time, it escalates to the program manager and steering committee.

An outline structure of the Program Management Office for both transition and transformation is shown in Fig. 7.12.

An outline of the overall governance structure is shown in Fig. 7.13.

FIGURE 7.12 PMO structure.

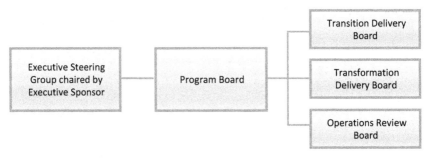

FIGURE 7.13 Governance.

Typical responsibilities and participants of these different governance bodies are outlined in Table 7.6.

7.6 CONCLUSIONS AND FUTURE OUTLOOK

So far, we discussed data, business processes, and data management technology. All this is meaningless without an organization that can act on these ingredients. Data is created and comes alive in the hands of users. The architectural best practices around data management need to be understood:

- There is an unambiguous definition of each data item. Together these form the enterprise data dictionary that is the organization's common vocabulary and the basis for any process.
- Data are shared, fit for purpose, accessible, secure and trustworthy, and compliant with regulations.
- The data management function is implemented on a minimum set of different technologies. Less is more in this case, and much less is much more. The smaller the number of copies of data in a certain domain, the better it is. Local copies are best avoided altogether.

Organizations that allow too many degrees of freedom in sources, definitions, and copies can never be a factory but will remain a wasteful workshop.

TABLE 7.6 Governance Components Responsibilities Overview

Group	Responsibility	Participants
Executive Steering Group	Responsible for successful outcome of the transition and transformation and: • Ensures relevant resources are available • Removes obstacles to progress • Reviews and provides oversight for the program's progress • Ensures business alignment • Reviews program risk and remediation plans • Approves key communications	Chaired by the client Executive Sponsor (responsible for setting goals, policy, and organizational alignment) Within the Steering Group are Executives representing client and service provider to whom exceptional escalations will be taken
Program Board	Responsible for the program including: • Program planning and execution • Risk Assessment and Planning • Program coordination across Functional Areas (IT, Operations, and Business) • Communication	Chaired by the Program Manager Client Program manager, Transformation Release Managers, Transition Manager, Operations Manager, risk, and key specialists
Transition Delivery Board	Responsible for executing the transition project	Chaired by the Program Manager Transformation Release Managers, Operations Manager, risk, project managers for areas in current and next release, and key specialists
Operations Review Board	Meets weekly. Responsible for the day-to-day management of the Operational processes. This includes client support for instrument data and any pricing queries	Chaired by Service Provider Operations Manager Client Operations management, Risk

Data has to be seen and treated as core business asset. Data owners are accountable for the quality of the data and the organization must define clear roles such as data stewards who are responsible for the quality of the data. Data governance policies have to go hand in hand with enterprise information management capabilities. Firms have put in place CDOs and data stewards to focus on good governance. The specific focus of a CDO on cost, revenue, or compliance may differ by firm.

Given the rise in data volumes and sources, and the arrival of technologies to use this data, the data management or data science function will increasingly be the source of competitive differentiation. Information businesses such as financial services will keep on shortening their cycles but are at the same time under heavy regulatory scrutiny. To properly navigate these challenges, no firm can afford the operational waste in redundant sourcing, the operational risk in making decisions on inaccurate, late, or incomplete data, and, above all, the opportunity cost of being behind the curve in understanding customer needs.

In information services as in content, the line between client and supplier is blurring. Increasingly, firms play both roles simultaneously. Organizations, technologies, and data management processes need to be ready for a data landscape of shared services and industry pooling models (N:M models). For the data in-house, the bar needs to be raised on governance, which starts with organizational discipline to abide by basic enterprise data architecture principles and to have data domain owners supported by clear decision-making processes that involve all stakeholders.

REFERENCES

Brose, M.S., Flood, M.D., Krishna, D., Nichols, B., 2014. Project implementation. Handbook of Financial Data and Risk Information, part II. Cambridge University Press, Cambridge, p. 425.

Churchman, C.W., 1967. Wicked problems. Manag. Sci. 14, 141–142.

Grody, A.D., Harmantzis, F., Kaple, G.J., 2006. Operational risk and reference data: exploring costs, capital requirements and risk mitigation. J. Operational Risk 1 (3.).

Ladley, J., 2012. Data Governance–How to Design, Deploy and Sustain an Effective Data Governance Program. Morgan Kaufmann, 10.

McGilvray, D., 2008. Executing Data Quality Projects–Ten Steps to Quality, Data and Trusted Information. Morgan Kaufmann.

Stockdale, D., 2014. Data governance and data stewardship. Handbook of Financial Data and Risk Information II. Cambridge University Press, Cambridge, pp. 480–481.

Chapter 8

What's Next?

Chapter Outline

8.1 STRUCTURAL CHANGES IN THE FINANCIAL SERVICES INDUSTRY

In this book, we have surveyed the information supply chain and the products and processes in the instrument and transaction life cycles. We have looked at data sets, business processes, regulatory needs, technology developments, data quality metrics, and organizational approaches to organize and monitor data flows. We have looked not only at data consumption, but also at new content production. Data becomes valuable when it is used. It is not just a *by-product* of doing business but a *corporate asset* in itself.

If data management processes are robust, the end result should be quality data fit for purpose. The complication is that the purpose can be different depending on the step in the transaction life cycle and specifics of a firm's business and the instrument type processed. Any application or user using data also brings with it certain assumptions and data quality can be in the eye of the beholder. In the information supply chain, the data model is critical as it is the foundation on which the software applications are going to be built. When designing information supply chains and data models, it pays off to keep the use cases in mind.

Whether data feeds are a hodgepodge of different elements or whether they can be standardized makes an enormous difference in sourcing information. Firms need to decide which of their content needs are industry needs, which are enterprise needs, and which are at division, department, or individual user level. This will inform data distribution and technology strategy and the sourcing options for data and data management. Outsourcing options can be considered only when an institution fully understands its information supply chain and understands what the KPIs are that go with their specific business needs.

A Primer in Financial Data Management. http://dx.doi.org/10.1016/B978-0-12-809776-2.00008-9
Copyright © 2017 Elsevier Ltd. All rights reserved.

We have discussed the need for horizontal roles across the traditional vertical siloed business lines. Simply put, siloed companies do not share best practices. The financial services sector is undergoing consolidation, data governance has so far always been a luxury in a merger. To prosper and possibly even to survive, the full potential of all data sourced and owned by the firm needs to be realized. For that, firms need to make sure that all data is accessible and connected. First of all, the meaning of data needs to be clear across all the users. Semantic data is smart data and with current business and regulatory requirements, firms cannot afford their data to be stupid. The meaning of data, feedback loops, and corrective mechanisms needs to be wired as low level as possible in data management processes.

However, companies today are typically caught in the data shallows meaning:

- They cannot tap into the data they have very quickly.
- They do not have much reach into their data silos.
- They cannot analyze their data very well.
- Their means for generating meaningful and high-impact insights is limited, infrequent, and immature.

Suboptimal access to data leads to data islands with corresponding cost, redundancy, and operational risk. The way to overcome the data shallows is clear data dictionaries and data governance processes.

The bar needs to be raised in data infrastructure given the general trends of:

- commoditization of financial products and reduction in risk taking by banks, and increased use of passive investment strategies;
- increased transparency for customers through pre- and posttrade transparency leading to more potential for price discovery and product comparison;
- use of data standards for reporting due to regulatory intervention and increased use of standard identification and classification standards;
- firms are moving to retain margins by finding new data "edges" by including nontraditional, unstructured data sources to get information advantages on customer behavior (results from rigorous analysis of short- and long-term data sets can challenge an organization's accepted wisdom and lead to new offerings and changes in client interaction);
- sourcing data more dynamically and interactively as customer information provision processes increasingly become real time;
- measuring and reporting on quality KPIs of information and feeding this information back continuously into the information supply chain.

Nobody ever has the luxury of setting up a complete greenfield infrastructure with the exception of newcomers. Incumbents are still heavily protected by regulation and capital requirements: the cost of an entry ticket into banking is high. Besides, client relationships—in particular retail clients—are normally sticky and switching costs and administrative burden can be high.

Regulation and the high cost of setting up an infrastructure used to be a very effective protective wall for financial services—a shield from start-ups and innovation. However, this is starting to change because

- Technology used to be a barrier to entry—now legacy technology in incumbent firms is a liability given the communication style, speed, and data access clients want. New firms can provide the differentiation, and customer experience via digital channels that existing financial services firms struggle to provide.
- The brands of financial services firms were badly damaged by the crisis.

8.2   THE SUPPLY CHAIN PERSPECTIVE OF INFORMATION MANAGEMENT

Financial institutions are information processing machines and based on the execution of business processes, new information is put back to clients, counterparties, investors and regulators. Since financial institutions are information generation enterprises, the cost and speed of pushing all raw information through to produce new content is the critical measure.

The information supply chain is a supply chain of intellectual property rather than physical property. Compensation for the owners of the goods is not always straightforward and usage of goods can sometimes be hard to determine. Also, contrary to physical goods, in the case of manufacturing information, the cycle time is very short and information products degrade very quickly in the financial space: like oysters they are best consumed fresh. The supply chain management challenges are therefore also different.

- Instead of a shipping date, timing can vary between microseconds for order latency and end of day for financial and risk reporting.
- The number of items to keep records on are millions of financial instruments, tens of thousands of professional counterparties, and millions of retail customers.

The supply chain is changing because raw data is captured closer to the source and many more nontraditional, unstructured data sources can be tapped and fed into decision making.

Tools and processes create a conveyor belt of information. Because the number of content suppliers is large and because parts of this supply chain will be serviced by third parties, there is an increased need for service level agreements (SLAs) with clearly defined service metrics. Less direct control over the various steps in the supply chain translates into a greater need for clear metrics to go into SLAs. Metrics as discussed in Chapter 6 may help us judge the effectiveness of the information supply chain in fulfilling the needs of its users that need to run the instrument and transaction lifecycle processes discussed in Chapter 4. They may help us define quality in terms that translate to business benefits for the users.

Client and regulatory demands have led to pressure on traditional data sourcing and curation processes. Firms should prevent locking data to specific workflows and applications but instead see data as an independent corporate asset. Historically however, career advancement in technology in financial services came from building new applications, not from improving or reusing existing infrastructure. Now there is an appreciation at last that information technology is more about building plumbing than about building cathedrals.

Taking away constraints to data access directly leads to productivity gains. An ideal situation would be a combination of a wiki plus enterprise search for all internal curated financial data. The future of information supply chains is that of a move toward piecemeal and ad hoc data creation and consumption. Increased application integration and breaking up of infrastructures into more flexible independent components will lead to more flexibility in sourcing and empower users through feedback loops quickly showing the effect of user feedback. On top of that, regulatory demands make that data lineage has to be wired into the supply chain.

Productivity and success measures for the information supply chain will be measured in different ways:

- The **back office** is about operational efficiency.Processes are set up for throughput and control and requires high-quality reference data and settlement data. The key KPI would be costs per transaction.
- The **front office** world is product, client, and ultimately opportunity driven. Since opportunities do not last, ideas need to be powered with new data sources fast. In product innovation we have the case of low "unit processing" high margin. You want to invest in ultralow-latency data, in price models, in quantitative models and in data scientists that would need access to cross-sectional data, and in securing exotic data not easily found via the standard aggregators. Disparate sources need to be linked together to find new patterns and to generate new product ideas.

These two processes translate into a "wholesale" and a "retail" flavor of information management. The retailer has to listen to end-user feedback and cater to local needs. The wholesaler has to source in large scale and is about B2B, the preordering of content in bulk at low per unit cost. The retail information management model needs to be nimble, B2C, and serve clients on-demand much like the local corner shop. In terms of different quality dimensions of data, the wholesale needs are in the areas of transparency, control, security, accuracy, synchronicity, and order. The retail needs are in the areas of relevance, completeness, accessibility, and speed. Consistency is equally important for both. The retailers also have to loop back and provide feedback to the wholesaler.

In addition, the data intensity of and scrutiny on risk and finance processes has gone up. Because risk and finance have to combine all the information flows from the business, they are hit hardest by ambiguity and data quality issues. Risk and finance themselves are merging since risk numbers have to produced more regularly and with greater regulatory scrutiny on the risk data aggregation processes. At the same time, valuation processes in accounting have moved in the direction of "mark to market" and have become more forward looking (e.g., in IFRS9 with the expected credit loss metrics; in short, credit impairments have to be taken as soon as possible taking different future scenarios into account).

A well-set-up information supply chain will create strategic options, not just through operational scalability to onboard new clients and create new products and services but also through options to decouple part of the information supply chain or the transaction lifecycle management. If an institution understands how its processes are implemented, understands the (meta)content needs of its businesses, and is capable to measure the operational effectiveness of them (through well-chosen and well-monitored KPIs), it is able to consider:

- outsourcing part of the information supply chain such as the integrating, consolidating, and validating of content, or part of the transaction life cycle including execution management, reporting, and asset servicing;
- replacing one third party in the transaction life cycle with another. Execution and clearing options will increase as increasingly the different segments in the transaction life cycle will be decoupled and serviced by specialist providers.

Apart from having the option to rearrange business processes, good data management also means that a firm will also be able to accurately *price* these various sourcing options as it will be aware of detailed baseline costs. Furthermore, if done well, this translates into the ability to use price differentiation for products and clients.

Insight into the data supply chain is a precondition to define metrics and to come to a Data Management Scorecard. The main considerations of a data management function include:

- Relevance: do we source all required data? What do we source redundantly? What is not needed?
- Timeliness: where are the waits on data availability? What are the bottlenecks?
- Cost-effectiveness: what do we spend on data? Can we slice and dice pockets of spend? Who is responsible for the largest data costs?
- Quality: what data requires rework? Where do quality issues arise? At source (manual, commercial source, client input) or at handover points in processes? How can we stem data quality issues at source?

Parts of the financial services industry are overcrowded and in the postcrisis world growth is lower which led to increased cost pressure. As far as information management is concerned, it seems obvious to centralize what you can centralize, internally and wherever possible across the industry. Centralization reduces operational and compliance costs and can also provide a foundation for more nimble processes. After all, if business applications do not have to worry about information needs, they can focus on value add and use those savings to invest in new capabilities. In addition to this, as the cost of technology falls, utilities are not only about cost reduction but also about an industry approach to solve industry problems by sharing risk and IP and coming up with joint solutions.

Financial services firms have an especially tough task innovating given the legacy of decades of building infrastructure, the constraints of regulation that fintech firms do not have, and the enormous amounts of applications and formats. Many firms are in the process of reducing the total number of applications and to open up remaining applications as much as possible. Financial services firms face a double challenge of making the existing operation more efficient while at the same time carving out channels of change. A bimodal model of legacy operation and innovation can be the answer: a traditional sourcing and curation process that can also be leapfrogged by injecting data from outside directly in the process and into the hands of end users.

Understandably, financial services firms seek collaboration with fintech suppliers who offer new approaches and faster cycle time in development. These collaborations can work very well given the client base and knowledge of regulation that financial services bring, combined with unorthodox approaches, faster cycling in development, and often a better understanding of client experience needs that fintech firms bring. To increase the speed of innovation, firms carve out channels of change and create sandbox environments to test innovation in specific processes or target client groups.

Apart from delivering large raw data sets to be mined, social media offers direct threats and opportunities to the classic intermediary role of financial services firms. Social media can help speed up disintermediation and replace financial services firms. Social media platforms are very good at mobilizing like-minded people and at building communities of interest. Communities such as LinkedIn interest groups or Meetup groups with shared business interests and/or geographies can be tapped to organize venture funding or to borrow money in a crowdsourcing model. Conversely, established financial service providers could offer tailor-made investment products for specific interest groups.

Banks and investment managers can embrace social media platforms and techniques to steer crowdfunding (the crowdfunding market is small but growing at a fast pace; estimated 2015 size is USD 34.4 billion; see http://www.crowdsourcing.org/editorial/global-crowdfunding-market-to-reach-344b-in-2015-predicts-massolutions-2015cf-industry-report/45376) and venture investments and for their own product development. Through social media, consumers self-segment and are doing the work of a marketing department for their suppliers.

For instance, part of the curation process of lending and investments could go through the review schemes common on sites such as Uber and AirBnB (we could expect more reviews of products on consumer sites, such as Morningstar's fund rating; from the regulatory side we already have seen the introduction of a standard Summary Risk Indicator in the EU). For any nontrivial investment amounts, it is likely that third parties are needed to screen and rate investments or lending and to certify and audit grassroots initiatives. These third parties could be banks or asset managers, but equally well they could be accounting firms or NGOs. Companies that tap into communities for microcredits or micro-investments could become new-style fund managers.

8.3 DATA MANAGEMENT OUTLOOK

A financial institution is its data. The financial services industry is recognizing that it needs to grow up in how it manages information as both the fundamental raw materials but also the end product of the business processes that it supplies to its customers, investors, regulators, and indeed its own staff. Industry issues need to be treated as such and should not be solved in-house by every individual firm. The recognition of industry data sets and enterprise-wide master data that needs to be easily accessible by all users and applications, supplemented with departmental or user-level, specific, and more sensitive data, is an essential part of data management.

Key themes here are *data liquidity* and *agility*: getting the information that is needed fast and at low cost. This requires substantial data integration, the breaking down of silos, the adoption of new techniques and the creation of new roles in a data management organization, a common data model, and the distribution into the hands of end users and business applications.

As we move from blind *processing* of information to more intelligent *managing* of information, more insight into the information supply chain will be created. Content products and information services will become more user friendly and will be better adapted to end-user workflow. Underlying data standards in terms of format but especially in terms of semantics and (ISO) standards on unique identification are a prerequisite. Financial institutions are increasingly recognizing the criticality of knowing and managing the information supply chain, not so much for regulatory and cost control reasons, but to establish a solid foundation for growth.

Existing business models can be a drag on innovation. There is always a fear of the new, a fear to jump in the dark as long as there is an existing franchise that is doing well. However, financial services firms realize that they risk being disintermediated and that they have to work to retain their position as trusted third parties. Partnerships with fintech firms are essential to provide improved customer service and to make the most of new ways of combining and analyzing information.

It is not possible to do solid and secure infrastructure on a shoestring: to attain a clean and clear 360-degree view of financial information and its supply chain, continuous investment is needed. The success of process improvement initiatives will be limited if they do not include enterprise content management.

Feedback loops where KPIs are fed back and acted upon upstream are essential. The key challenge for any information supply chain is how to simultaneously resolve data needs, for example, for STP, cost reduction, and system rationalization while at the same time providing added value for customers.

How do we consume and select financial products in the future? Improved data management can lead to both more transparent and more tailored financial products that help people be optimally safe and insure, borrow, and invest according to their needs and preferences. The blending of data and business logic in smart contracts makes that not only terms and conditions but also events contingent on certain external data values are included (such as the value of an index or creditworthiness of a certain party). The data contains its own triggers and this can lead to new instruments with automated actions or payoffs based on external triggers and reference points. (We have seen this, e.g., in the case of the low-cost Robo-advisors in wealth management with automated asset allocation based on investor risk appetite. Triggers and timing can be linked both to market movements and to life events such as retirement, education funding, and so on.)

In an industry as data intensive as financial services, every employee is necessarily a data manager. However, the key constraint on productivity for knowledge workers is timely access to good information. Staff that is less constrained by data access and that can focus on producing new information instead of wasting time on hunting for or double checking information will be vastly more valuable to companies.

Improved information supply chains are the basis for streamlined and scalable operations. At the same time, advances in feeding the human mind with information will lead to direct competitive advantage as it will free up staff to come with new offerings and focus on client interaction. The typical competitive model looks at either low cost or differentiation as competitive advantages a firm can have. Sound data management is a precondition for both.

Bibliography

Recommended books for further reading include the following:

Adelman, S., Moss, L., Abai, M., 2005. Data Strategy. Addison Wesley, Upper Saddle River, NJ.

Alvarez, M., 2007. Market Data Explained: A Practical Guide to Global Capital Markets Information. Elsevier and Mondo Visione World Capital Markets, Burlington, MA.

Brose, M.S., Flood, M.D., Krishna, D., Nichols, B. (Eds.), 2014a. Handbook of Financial Data and Risk Information I: Volume 1: Principles and Context. Cambridge University Press, Cambridge, UK.

Brose, M.S., Flood, M.D., Krishna, D., Nichols, B. (Eds.), 2014b. Handbook of Financial Data and Risk Information II: Volume 2: Software and Data. Cambridge University Press, Cambridge, UK.

Freedman, R.S., 2006. Introduction to Financial Technology. Elsevier, Boston, MA.

Gordon, K., 2013. Principles of Data Management—Facilitating Information Sharing, second ed. BCS, The Chartered Institute for IT, Swindon.

Kaplan, J.M., Bailey, T., O'Halloran, D., Marcus, A., Rezek, C., 2015. Beyond Cybersecurity. Wiley, Hoboken, NJ.

King, B., 2012. Bank 3.0. Marshall Cavendish Business, Singapore.

Ladley, J., 2012. Data Governance—How to Design, Deploy and Sustain an Effective Data Governance Program. Morgan Kaufmann, Waltham, MA.

McGilvray, D., 2008. Executing Data Quality Projects—Ten Steps to Quality, Data and Trusted Information. Morgan Kaufmann, Burlington, MA.

Olson, J.E., 2003. Data Quality—The Accuracy Dimension. Morgan Kaufmann, San Francisco.

Redman, T.C., 2008. Data Driven. Harvard Business Press, Boston, MA.

A Primer in Financial Data Management. http://dx.doi.org/10.1016/B978-0-12-809776-2.00012-0
Copyright © 2017 Elsevier Ltd. All rights reserved.

Index

Printed in the United States
By Bookmasters